Landscape Guide

The South Central States

by Tom Clote, Jr.

Published by Landscape Guides, LLC

First Edition
Cover Design by Chris Pearl

Acknowledgements

My thanks to Dr. T. H. Milby, University of Oklahoma Retired Professor of Botany, who generously reviewed the area plant lists; to Craig Evans, Cleveland County, Oklahoma extension agent, who wrote the section on water and proved a helpful consultant; to Steve Hill, MLA, Adjunct Professor of Landscape Architecture at the University of Oklahoma, who kindly contributed numerous slides used in the publication, and to Martha Jennings, Travis Caperton, Andrea Hahm, and Pat Foley who also contributed slides.

Thanks also to Carolyn Simpson, who artfully crafted something that is easy to read, and to Joe Clote, Carlos Amaya, Brad Taylor, and Sandy Bent.

Most of all I would like to express my gratitude to my wife, Sallie, and my son, Alex, whose love, patience, and support made this book possible.

Sources for business listings, arboretums, and botanical gardens include the department of agriculture for the states of Texas, Oklahoma, Arkansas, and Louisiana. Other sources include yellow page directory listings from these respective states.

Every effort has been made to make sure that all information is accurate. However, it is not possible to guarantee total accuracy.

Table of Contents

Introduction

HOW THE LANDSCAPE GUIDE CAN HELP YOU

I decided to write a landscaping book after listening to the same old complaints from friends and clients who were mostly weekend gardeners. They groused about wasting a whole day, driving from store to store trying to find a certain item only to be disappointed upon arrival. In an industry with few standardized products, selecting the right store and the right plants takes a lot of legwork and know-how.

Friends would ask my advice about which plants to buy or where to get the best deal, not realizing that selecting the right plants and trees is more complicated than that. With that in mind, I decided to review over 1600 nurseries, garden centers and sod farms in this 4 state region to show you-the consumer-where they are and what they might be able to do for you. To the extent that the various nurseries or garden centers replied to my questionnaires, I have addressed a business' reputation, years of operation, quantities of native plants, in-house landscape design services, and availability of organic pesticides and fertilizers. Abbreviations are used for easy identification of business services. Nurseries, garden centers, and sod farms are color coded by area. All listings include the business name, location, telephone numbers, and hours of operation, as available. Some stores are especially unique, and I will underline their specialty.

Additionally, I've included some general advice about soil preparation, watering your plants, designing a garden space (utilizing style, structure and color), and how to transplant your plants from the nurseries to your garden space. My objective is to make the landscaper's knowledge more accessible to you, the weekend gardener.

Area Plant Communities

The Four-State Area

The plant communities of Texas, Oklahoma, Arkansas, and Louisiana are very diverse, but each broad area shares many qualities in common. Obviously you'll want to consider your region's native plants before making your landscape plans in order to know best what will thrive there. The region's native plants reveal soil types, water requirements, and heat and cold tolerances.

Plant communities are instructive in another sense: despite being diverse, they are interdependent. For example, locust and mesquites are legumes. They convert nitrogen from the atmosphere and help put it in the soil. Nitrogen is vital to the development of plant life.

Pines and deciduous softwoods in their natural environment help stabilize the soil and provide a more protective environment for oaks and other hardwoods. Black walnut and artemesia (wormwood) have certain properties that inhibit the growth of seedlings and several plant species. The leaves and wood of juniper contain a resin (sandarac) which is obnoxious to many insects. Black locust bark has a carbohydrate containing protein that causes the clumping of red blood cells. These are toxic to some insects. Following this section are lists of plants by area, an asterick indicates which plants are native. While not all native plants are available or even practical for your garden, you can make suitable substitutions if you're knowledgeable. Let's consider some of the native plants in the South-Central United States.

Ozarks and Ouachita Mountains

Ash, buckeye, elm, hackberry, maples, oaks, plums, wild cherry, dogwood, redbud, and holly are among the many tree species found in the Ozarks and the Ouachita mountains of Arkansas and eastern Oklahoma. American bellflowers, bluets, spring beauty, wild verbenas, orchids, ferns, and herbs all thrive under the favorable growing conditions of these mountains. The soil of the Ozarks is typically rocky and loamy. The Ouachita's soil tends to be silty and sandy, but more fertile soil is found in the river valleys of these areas. Due to the nature of the mountainous terrain with the many crevasses and different rock protrusions, various microenvironments allow a variety of distinct plant populations.

A lazy river near the Ouachita Mountains.

Eastern Texas and Western Louisiana

The piney woods of eastern Texas and western Louisiana have a loamy or sandy loam soil with plants that are adapted to very wet conditions but tolerate dry as well. The native longleaf pine has mostly been supplanted by the shortleaf and loblolly pine. Mixed with the pine woods are red and white oak with a smattering of dogwoods, sassafras, ferns, herbs, and wildflowers.

Hardwood stands of oak, hickory, water tupelo, cypress and gum grow in lowland areas such as the Sabine River and Ouachita River basins. They support a variety of ferns, wildflowers, orchids, sedges, and grasses. The lowland soil is mostly heavy clay, which becomes very "mucky" when wet. The bald cypress tree has developed knees or breathing roots (pneumatophores) in order to adapt to mucky (wet and sticky) soil.

The 'Piney Woods' of Western Louisiana.

The Mississippi Alluvial Plain

The Mississippi Alluvial Plain runs its course along the river from Northern Arkansas on down to north of Lake Ponchatrain in Louisiana. The benefits of past and present courses of the river have left a deposit of rich alluvial (river bottom) soil. Hardwood stands of oak, hickory, tupelo, gum, and bald cypress thrive here. Ridges along the plain in Louisiana are dominated by live oak.

Coastal Louisiana and Texas

The coastal prairies and marshes extend along the coasts of Louisiana and Texas. The coastal prairie used to be an extensive tall grass prairie, extending from the piney woods of east Texas and Louisiana to the marshes along the shoreline. The native tall grass prairie is similar in composition to the Midwest prairie except for the vast amount of water. Plants in this region have adapted to water logged conditions.

Soils of the coastal prairie tend to be wet, heavy, and clayey. Peat and muck soils are found in the coastal marshes. Due to the influx of the Gulf waters, and decreased freshwater flows, there is a constant threat of saltwater intrusion along the coastal marshes.

Thriving plants can tell you how much salt is in the water. The salt marsh has smooth cordgrass, the brackish marsh has marsh-hay cordgrass, and the freshwater marshes have marsh hay, cattails, and sawgrass. Black-mangrove and marsh/scrub are found along the gulf coast of Louisiana and east Texas. Cypress, water tupelo, willows, swamp privet, American and yaupon holly thrive in the swamps and coastal plains.

Coastal Marsh at Sabine Wildlife Refuge, La.

Southeast Texas Understory

East Central Texas and Oklahoma

The post oak belt runs from east central Texas northwards through east central Oklahoma. The predominant mix of trees are post and black jack oak fringed by cedars, redbuds and sumac. The soils alternate from sandy loam to clay. Occasionally to the east, but primarily to the west of the post oak belt lie the tall grass prairies, containing big bluestem grass, Indian grass, switch grass and others. Wild indigos, goldenrods, anemones, sunflowers, verbenas, violets, golden coreopsis, primroses, blazing star and butterfly weed make up only a small part of the prairie flowers in this area. The plants that have adapted to this area must be tolerant of long periods of drought.

Post Oak & Black Jack Oak Belt

The Texas Hill Country

The Texas Hill Country (Edwards plateau) features limestone cliffs, granite outcrops, springs and rivers. Ash, juniper, black hickory, Texas madrone, lacey oak, escarpment live oak, and mesquite are all native trees to this area. Wildflowers come alive every spring and include: bluebonnets, Indian paintbrush, Indian blanket, Texas star daisy, yarrow poppies, daisies, lemon mint and pink evening primrose. Soils in this region tend to be well drained alkaline soils. The plant life of this region has adapted to periods of prolonged droughts and flash floods.

Indian paint brush and bluebonnets in the Hill Country.

The High Plains of Texas and Oklahoma Panhandles

The high plains of the Texas and Oklahoma panhandles and northwest Oklahoma (the shortgrass prairie) have the short grasses: little bluestem, buffalograss, and blue gamma with sage, cactus, wildflowers such as penstemon, plains melampodium, sunflowers, plains wallflower and an occasional stand of shinnery oak. These natives do well despite scant rainfall and a short growing season.

Gallardia, Monarda, Coreopsis, Artemesia, Yucca and Solanum grow well in Roger Mills County, Oklahoma.

West Texas and the Trans Pecos

West Texas and the trans Pecos (west of the Pecos River) is an area of mesquite, yucca, tarbush, creosote bush, snakeweed, saltbush and cactus. The soil is characteristically fine textured heavy soil. The wind here is harsh, in addition to the hot temperatures. Plants do respond to water, and they are especially vibrant and colorful after heavy rain.

Prickly pear cactus, candelilla, and Texas sage in Big Bend National Park.

South Texas

South Texas (mesquite brushland) has a diversity of species of mesquite and cactus. The roots of plants here are well developed in adapting to the hot and dry conditions. People say the mesquite tree's growth above ground is like the tip of an iceberg. Aboveground plants have developed small leaves in order that they might better survive. Plants that survive here must have a well- developed root system, and small leaves in order to adapt to the hot and dry conditions.

Agave, Pincushion and Barrel Cactus

Climate Zone Map

This map is produced by the USDA. It is used in nursery catalogs and garden books. The zones are an average low temperature for winter temperatures with a 10 degree range.

Find out where you reside on the zone map and take note that that is the zone for plants in your area.

Temperature Zones 6 through 9

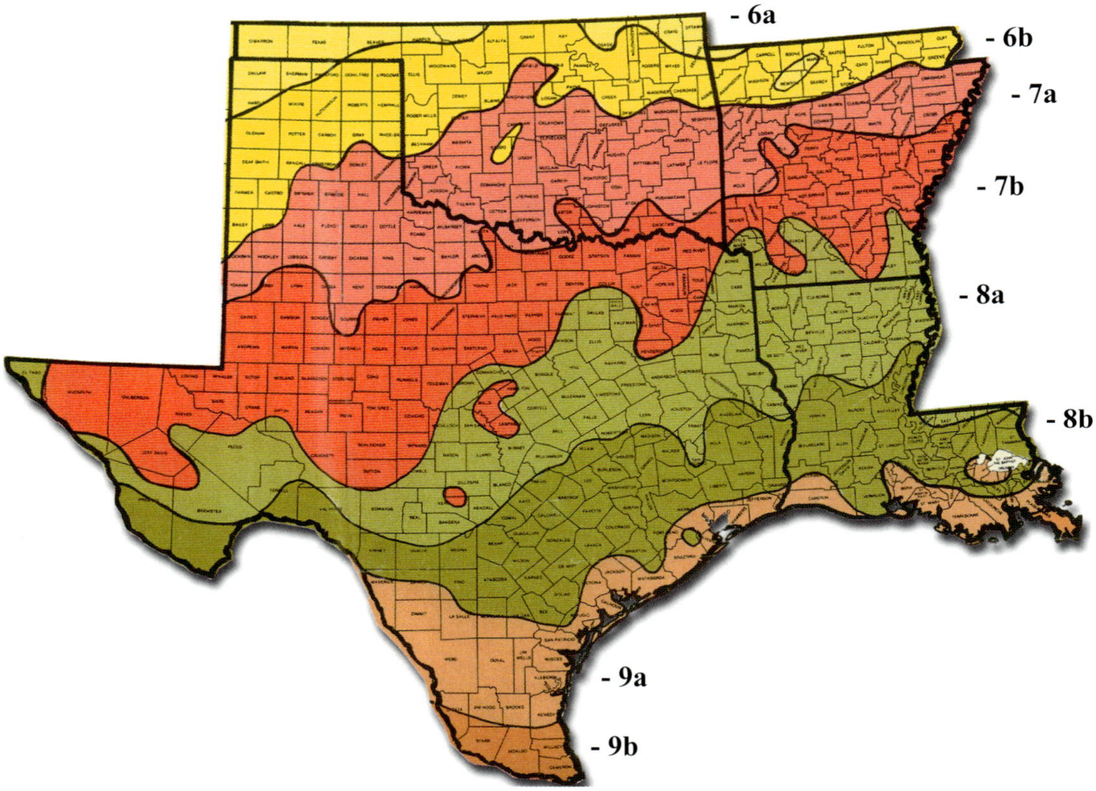

Texas, Oklahoma, Arkansas, and Louisiana have four hardiness zones.

Each zone is based on a 10 degree Fahrenheit difference in average annual minimum temperature.

USDA ZONE	Temperature Range	Sample City
6a	-10 to -5 F	Dalhart, Texas
6b	-5 to 0 F	Tulsa
7a	0 to 5 F	Oklahoma City
7b	5 to 10 F	Little Rock
8a	10 to 15 F	Dallas
8b	15 to 20 F	Austin
9a	20 to 25 F	New Orleans
9b	25 to 30 F	Brownsville, Texas

Rainfall Averages Map

Observe the dramatic shift of rainfall in transition from east to west. Plants are greatly impacted by these patterns and the appearance of plants is one from lush to stark. Careful observation of native flora reveals the conditions plants have adapted to.

Plant with the understanding that the rainfall figures provided are averages. Some years have rainfall totals far more generous than others.

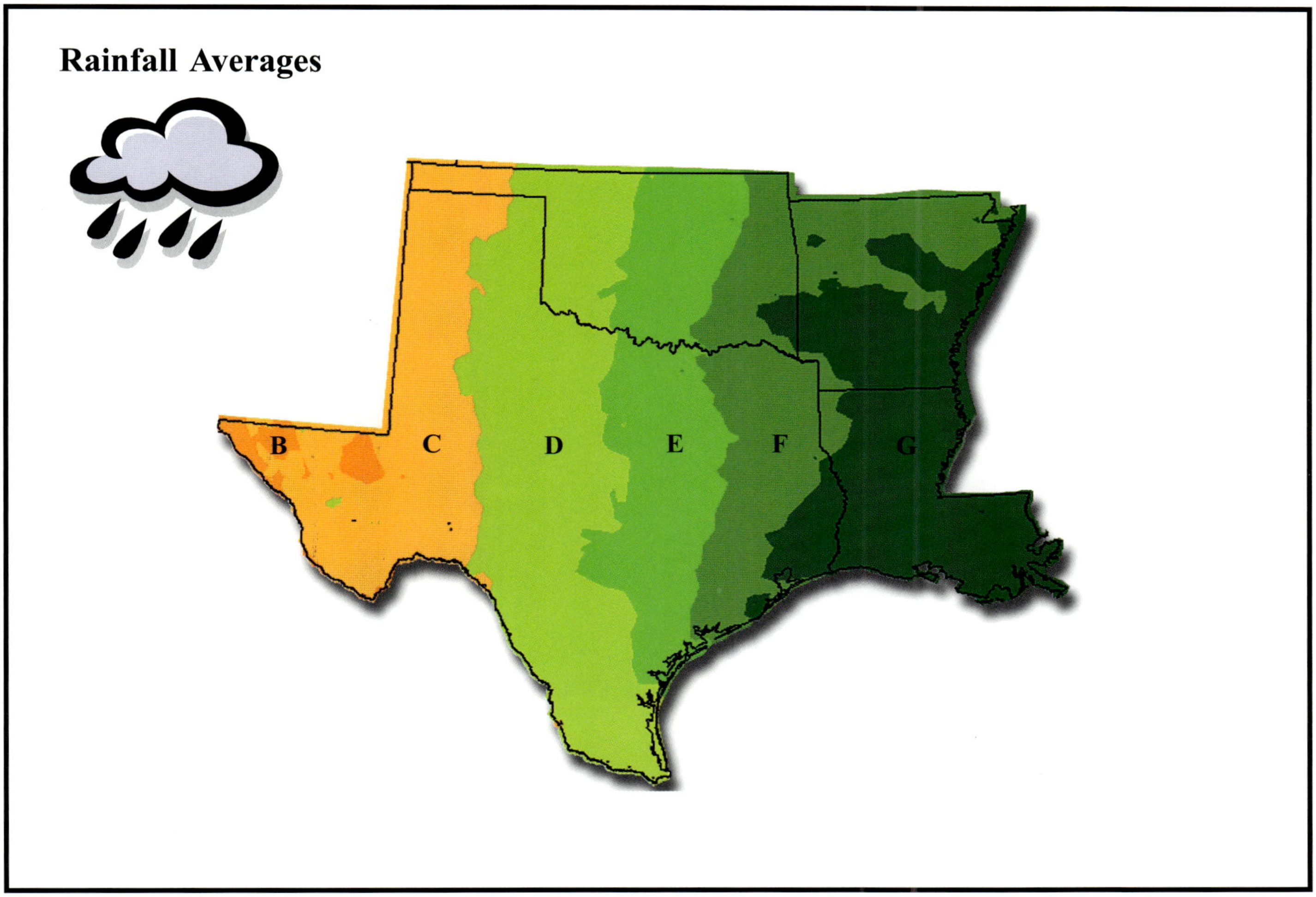

Map is courtesy of climate atlas of the United States.
Published by the National Climatic Data Center in Ashville, N.C.
Date of publication September 2000

B	5 to 12 inches
C	12 to 20 inches
D	20 to 30 inches
E	30 to 40 inches
F	40 to 50 inches
G	50 to 60 inches

Area Plant Lists

The plants listed for each area have been carefully considered. Plants are not exclusive to one list and they are in many instances interchangeable with nearby areas. Five requirements to consider when planting are: 1) Plant hardiness zones (refer to zone map on page 8); this sometimes varies locally within an isolated area, 2) Watering needs, 3) Soil requirements, 4) Sunlight needs, and 5) Design elements.

Th popularity of using native plants has grown enormously in the decade of the 1990's. It is no mere coincidence that the same decade experienced tremendous urban development. The use of natives helps reinforce identity with the displaced environment.

A common misconception is that all natives are easy to grow. The Longleaf pine at the 'grass stage', it's early years, is slow and difficult at best to grow. Others like the Western Soapberry are easily transplanted and adapt well. Remember, select the right plant for 'that' location.

Northern & Central Arkansas **(Fayetteville, Fort Smith, Little Rock)**

**denotes native to the 4-state area*

Trees:

American Beech *Fagus grandifolia* *
Chanticleer Pear *Pyrus calleryana 'Chanticleer'*
Bald Cypress *Taxodium distichum* *
Cherry Laurel *Prunus caroliniana* *
Chinese Pistache *Pistachio chinesis*
Chinkapin Oak *Quercus muhlenbergia* *
Crabapple *Mallus sp.*
Dawn Redwood *Metasequoia glytostroboides*
Downy Serviceberry *Amelanchier canadensis*
Eastern White Pine *Pinus strobus*
Flowering Dogwood *Cornus florida* *
Fringe Tree *Chionanthus virginicus*
Ginkgo *Ginkgo biloba*
Golden Raintree *Koelreuteria paniculata*
Japanese Dogwood *Cornus kousa*
Japanese Maple *Acer palmatum*
Japanese Red Pine *Pinus densiflora*
Leland Cypress *Cupressocyparis leylandii*
Loblolly Pine *Pinus elliotti* *
Pecan *Carya illinoensis* *
Pin Oak *Quercus palustris* *
Redbud *Cersis canadensis* *
River Birch *Betula nigra* *
Saucer Magnolia *Magnolia soulangiana* *
Scotch Pine *Pinus sylvestris*
Southern Magnolia *Magnolia grandiflora* *
Sugar Maple *Acer saccharum* *
Sweetgum *Liquidambar styraciflua* *
Sycamore *Platanus occidentalis* *
Tulip Poplar *Liriodendron tulipifera* *
Washington Hawthorn *Crataegus phaenopyrum*
Yaupon Holly *Ilex vomitaria* *

Shrubs:

Althea *Hibiscus syriacus*
American Holly *Ilex opaca* *
Barberry *Berberis thunbergii*
Boxwood *Buxus microphylla*
Burning Bush (Dwarf) *Euonymous altus 'compacta'*
Chinese Holly *Ilex cornuta*
Cleyera *Cleyera japonica*
Crape Myrtle *Lagerstroemia indica*

Snowball Bush

Beech Tree

American Holly

Forest Pansy Redbud

Exbury Azalea *Rhododendron exbury* *
Forsythia *Forsythia x intermedia*
Glossy Abelia *Abelia Grandiflora*
Heavenly Bamboo *Nandina domestica*
Hydrangea *Hydrangea macrophylla*
Japanese Boxwood *Buxus japonica*
Japanese Kerrai *Kerria japonica*
Japanese Skimmia *Skimmia japonica*
Juniper *Juniperus sp.*
Kurume Azalea *Rhododendron obtusum*
Leatherleaf Mahonia *Mahonia bealei*
Leatherleaf Viburnum *Viburnum rhytidophyllum* *
Lilac *Syringa vulgaris*
Manhattan Euonymous *Euonymous Kiautschovicua*
Mountain Boxwood *Pachistima canaby*
Mountain Laurel *Kalmia latifolia* *
Nandina *Nandina domestica*
Oregon Grapeholly *Mahonia aquifolium*
Rockspray Cotoneaster *Cotoneaster horizontalis*
Rhododendrons *Rhododendrons hybrids*
Rose Daphnes *Daphne cneorum*
Virginia Sweetspire *Itea virginica* *
Snowball Bush *Viburnum sp.*
Star Magnolia *Magnolia stellata* *
Yew *Taxus x media*

Emerging Cinnamon Fern

Daylillies

Flowers:

Ageratum *Eupatorium coelestinum*
Astilbe *Astilbe sp.*
Basket-of-Gold *Aurinia saxatilis*
Bearded Iris *Iris xiphiodes*
Blanket Flower *Gaillardia x pulchella* *
Bloodroot *Sanguinaria canadensis*
Candytuft *Iberis sempervirens*
Cardinal Flower *Lobelia cardinalis* *
Creeping Phlox *Phlox sublata*
Confederate Rose *Hibiscus mutabilis* *
Coral Bell *Heuchera*
Garden Canna *Canna x generalis*
Gaura *Gaura lindheimeri* *
Daylillies *Hemerocallis sp.*
Flag Iris *Iris pseudocorus*
Gayfeather *Liatris pycnostachya* *
Indian Blanket *Guillardia sp.*
Japanese Iris *Iris kaemferi*
Lantana *Lantana sp.* *
Lily *Lilium sp.*
Lungworts *Pulmonaria sp.*
Meadow Anemone *Anemone canadensis* *
Moonshine Yarrow *Achillea x moonshine* *
New England Aster *Aster novai-angliae*
Obedient Plant *Physostegia virginiana* *
Sunray Coreopsis *Coreopsis grandiflora*
Sweet Violet *Viola odorata*
Tickseed *Coreopsis grandiflora* *
Verbena *Verbena sp.* *
Veronica (Speedwell) *Veronica sp.*
Wild Ginger *Asarum sp.* *
Yellow Cordalis *Corydalis lutesa*

Bearded Iris

Carolina Jasmine

Creeping Phlox

Vines, Ornamental Grasses, and Groundcovers:

Bittersweet Vine *Celastrus scandens*
Coral Honeysuckle *Lonicera sempervirens* *
Carolina Jasmine *Gelsemium sempervirens* *
Clematis Vine *Clematis sp.* *
Lilly-of-the-Valley *Convallaria*
Moneywort *Lysimachia nummularia*
Monkeygrass *Liriope muscari*
Periwinkle *Vinca minor*
Pachysandra *Pachysandra terminalis*
Roses Climbers and Carpet *Rosa sp.*
Soloman's Seal *Polygonatum biflorum*
Trumpet Vine *Campsis radicans* *
Virginia Creeper *Parthenocissus quinquefolia* *
Wild Strawberry *Waldsteinia*
Winter Creeper *Euonymous fortunei*
Wisteria *Wisteria sp.*
Yellow Archangel *Lamiastrum galeobdolon* *

Central and Southern Louisiana

(Lake Charles, Lafayette, Baton Rouge, New Orleans)

denotes native to the 4-state area

Trees:

Black Gum *Nyssa sylvatica* *
Citrus *Citrus sp.*
Common Persimmon *Diospyros virginiana* *
Devil's Walking Stick *Aralia spinosa* *
Flowering Dogwood *Cornus florida* *
Fringetree *Chionanthus virginicus* *
Laurier Amande (Cherry Laurel) *Prunus caroliniana* *
Live Oak *Quercus virginiana* *
Loquat *Eriobotrya japonica*
Nutall Oak *Quercus nuttalli* *
Red Bay *Perseaborbonia* *
Sassafras *Sassafras albidum* *
Slash Pine *Pinus elliottii* *
Sourwood *Oxydendrum arboreum* *
Southern Magnolia *Magnolia grandiflora* *
Southern Red Oak *Quercus falcata* *
Spruce Pine *Pinus glauba* *
Strawberry Tree *Arbutus unedo* *
Sweetbay Magnolia *Magnolia virginiana* *
Taiwan Cherry *Prunus campanulata*
Titi Cliftonia *Monophylla*
Winged Elm *Ulmus alata* *
Yellow-Popular *Liriodendron tulipifera* *

Black Gum Tree

Yellow Poplar Tree

Shrubs:

American Beautyberry *Callicarpa americana* *
Aucuba *Aucuba japonica*
Azalea *Rhododendron sp.*
Beauty Berry *Callicarpa americana*
Bottlebrush *Callistemon lanceolatus*
Camellia *Camellia japonica*
Crape Myrtle (dwarf) *Lagerstroemia indica nana*
Fatsia *Fatsia japonica*
Fetter Bush *Lyonia lucida* *
Florida Anise (Starbush) *Illicium floridanum**
Gardenia *Gardenia jasminoides*
Hibiscus *Hibiscus rosa-sinesis*
Indian Hawthorn *Raphiolepsis sp.*
Inkberry (Winterberry) *Ilex glabra* *
Ixora *Ixora coccinea*
Japanese Yew *Podocarpus macrophyllus*
Laurustinus *Viburnum Tinus*
Mexican Orange *Choisyaternata*
Nandina *Nandina domestica*
Natal Plum *Carissa grandiflora nana**
Oleander *Nerium oleander*
Pineapple Guava *Feijoa sellowiana*
Pittisporum *Pittisporum tobira*
Rabbit-Eye-Blueberry *Viccinium ashei* *
Sasanqua Camellia *Camellia sasanqua*
Strawberry Bush *Euonymous americana* *
Sweet Olive *Osmanthus fragrans* *
Virginia Willow *Itea virginica* *
Wax Myrtle *Myricacerifera* *
Witch Hazel *Hamamelis virginiana* *
Yaupon *Ilex vomitoria* *

Horsetail

American Beautyberry

Ferns:

Cinnamon Fern *Osmunda regalis*
Holly Fern *Cyrtomium falcatum*
Ebony Spleenwort *Asplenium platyneuron* *
Leatherleaf Fern *Rumohra adiantiformis*
Mariana Maiden Fern *Thelypteris torresiana*
Sword Fern *Polystichum munitum*
Variegated Shield *Archniodes simplicior*

Stokes Aster

Flowers:

Agapanthus (Blue Lily of the Nile) *Agapanthus africanus*
Amaryllis *Amarillus hybrids*
Bear's-Breeches *Acanthus spinosus*
Bog Lily *Crinum americanum* *
Bog Sage *Salvia uliginosa* *
Brazilian Verbena *Verbena bonariensis*
Canna *Canna sp.*
Chameleon Plant *Hymenocallis caroliniana*
Daylily *Hemerocallis*
Four O'clock *Mirabilis jalapa**
Goldenrod *Solidago sp.* *
Jacobinia *Justicia carnea*
Louisiana Blue Phlox *Phlox divaricata* *
Louisiana Iris *Iris fulva* *
Marguerite *Chrysanthemum frutescens*
Mexican Marigold *Tagetes lucida*
Mist Flower *Eupatorium*
Milkweed *Asclepias sp.* *
Nun's Orchid *Phaius grandifolius* *
Plumbago *Plumbago ariculata*
Purple Heart *Secreasea pallida 'Purple Heart'*
Shrimp Plant *Justicia brandegeana*
Spider Lily *Hymenocallis*
Swamp Hibiscus *Hibiscus militaris* *
Swamp Lily *Hymenocallis caroliniana*
Tall Phlox *Phlox descussata*
Tuberose *Polianthes tuberosa*

Four O'Clock

Monkey Grass

Tall Phlox

Groundcover, Vines, and Grasses:

Algerian Ivy *Hedera canariensis*
Angel-wing Jasmine *Jasminum nitidum*
Asparagus Fern *Asparagus sprengeri*
Cordgrass *Spartina patens* *
Confederate Jasmine *Trachelospermum jasminoides*
Baby's Tears *Soleirolia soleirolii*
Carolina Jasmine Vine *Gelsemium sempervirens* *
Coral Vine *Antigonon leptopus*
Cross Vine *Bignonia capreolata*
Evergreen Miscanthus Grass *Miscanthus transmorrisonensis*
Evergreen Wisteria Vine *Millettia reticulata*
Fall Ruby Grass *Rycheytrum neriglume*
Feather Reed Grass *Briza media*
Kiwi Fruit Vine *Actinidia chinesis*
Monkey Grass *Liriope*
Pepper Vine *Ampelopsis arborea*
Swamp Jessamine Vine *Gelsemium rankinii*
Trailing Lantana *Lantana montevidensis*
Wedelia *Wedelia tribolata*
Weeping Lovegrass *Eragrostis curvula*

Wood Fern

Southwestern, Central, and Eastern Oklahoma

(Lawton, Oklahoma City, Tulsa)

**denotes native to the 4-state area*

Trees:

Amur Maple *Acer ginnala* *
Bald Cypress *Taxodium distichum* *
Blue Atlas Cedar *Cedrus atlantica*
Burr Oak *Quercus macrocarpa* *
Caddo Sugar Maple *Acer saccharum 'Caddo'* *
Cedar of Lebanon *Cedrus libani*
Cherry Laurel *Prunus caroliniana* *
Chinese Pistache *Pistachio chinesis*
Chinkapin Oak *Quercus muhlenbergia* *
Chittamwood *Bumelia lanuginosa* *
Ginkgo *Ginkgo biloba*
Japanese Maple *Acer palmatum 'Crimson Queen'*
Kentucky Coffee Tree *Gymnocladus dioica* *
Kwanzan cherry *Prunus serrulata*
Lacebark Elm *Ulmus parvifolia*
Lacebark Pine *Pinus bungeana*
Loblolly Pine *Pinus taeda* *
Pecan *Carya illinoensis* *
Ponderosa Pine *Pinus ponderosa* *
Redbud *Cersis canadensis* *
River Birch *Betula nigra* *
Sand Plum *Prunus angustifolia* *
Sawtooth Oak *Quercus acutissima* *
Scotch Pine *Pinus sylvestris*
Shumard Oak *Quercus shumardii* *
Smoke Tree *Cotinus coggygria* *
Southern Magnolia *Magnolia grandifolia 'Edith Bogue'* *
Sycamore *Platanus occidentallis* *
Tanyosho Pine *Pinus densiflora 'Umbraculifera'*
Washington Hawthorn *Crataegus phaenopyrum*
Western Soapberry *Sapindus drumondii* *
Yaupon Holly *Ilex vomitaria* *

Hydrangea

Caddo Maple

Washington Hawthorne

Shrubs:

Althea *Hibiscus syriacus*
Bayberry *Myrica cerifera*
Burning Bush *Eonymous altus 'compacta'*
Butterfly Bush *Buddleia davidii*
Chaste Tree *Vitex*
Chinese Holly *Ilex cornuta*
Coralberry (Indian Current) *Symphoricarpos orbiculatusu* *
Crape myrtle *Lagerstroemia indica*
Eleagnus *Eleagnus fruitlandi*
Frazer Photinia *Photinia x frazeri*
Flowering Quince *Chaenomeles japonica*
Forsythia *Forsythia sp.*
Glossy Abelia *Abelia Grandiflora*
Japanese Boxwood *Buxus japonica*
Juniper *Juniperus sp.*
Hydrangea *Hydrangea macrophylla*
Lavender Tree (Chaste Bush) *Vitex agnus-castus**
Leatherleaf Mahonia *Mahonia bealei*
Mugo Pine *Pinus mugo*
Nandina *Nandina domestica*
Oakleaf Hydrangea *Hydrangea quercifolia* *
Ozark Witch-hazel *Hamamelis vernalis* *
Purple-leaf Sandcherry *Prunus x cistena* *
Rusty Blackhaw *Viburnum rufidulum* *
Slender Lespedeza *Lespedeza virginica* *
St. John's-wort *Hypericaceae* *

Lacebark Pine

Chaste Tree

Warren Red Possumhaw *Ilex decidua 'Warren Red'* *
Wintergreen Boxwood *Buxus microphylla koreana*
Snowball bush *Viburnum sp.*
Virginia Sweetspire *Itea virginica* *
Yaupon Holly (dwarf) *Ilex vomitoria 'nana'* *

Lilly of the Valley

Flowers and Herbs:

Autumn Sage *Salvia greggii* *
Baldwin Ironweed *Veronia baldwini* *
Bearded Iris *Iris xiphiodes*
Bellflower *Cananula carpatica*
Bleeding Hearts *Dicentra spectabillis*
Blue Wild Indigo *Baptisia australis* *
Candytuft *Iberis sempirens*
Canna *Canna x generalis*
Columbine *Aquilegia hybrids* *
Coral bells *Heuchera*
Daylilly *Hemerocallus sp.*
Evening Primrose *Oenothera sp.* *
Four O'clock *Mirabilis jalapa* *
Garden Canna *Canna x generalis*
Gaura *Gaura lindheimeri* *
Hardy Hibiscus *Hibiscus moscheutos* *
Hollyhocks *Alcea rosea*
Indian Blanket *Guillardia sp.* *
Lantana *Lantana sp.* *
Maximillian Sunflower *Helianthus maximiliani* *
Meadowrue *Thalictrum dasycarpum* *
Ox-eye Daisy *Chrysanthemum leucanthemum**
Peony *Peony lactiflora*
Plains Larkspur *Delphinium virescens* *
Purple Coneflower *Echinacea sanguinea* *
Pincushion Flower *Scabosa columbaria*
Red Hot Poker *Kniphofia uvaria*
Silver King Artemesia *Artemesia ludoviciana**
Snow-on-the-Prarie *Euphorbia bicolor* *
Speedwell *Veronica*
Spiderwort *Tradescantia ohiensis* *
Sunray Coreopsis *Coreopsis grandiflora*
Sweet Violet *Viola odorata*
Winecup *Callirhoe involucrata* *
Wood Sorrel *Oxalis crassipes*
Yarrow *Achillea millefolium* *

Columbine

Maximillian Sunflower

Indian Blanket

Vines, Ornamental Grasses, and Groundcovers:

Big Bluestem *Andropogon gerardi* *
Blue-Gray Sedge *Carex glauca* *
Carpet and Climbing Rose *Rosa sp.*
Coral Honeysuckle Vine *Lonicera sempervirens* *
Carolina Jasmine Vine *Gelsemium sempervirens* *
English Ivy *Hedera Helix*
False Strawberry *Waldsteinia fragariodes*
Fountain Grass *Pennistum*
Hosta *Hosta sp.*
Indian Grass *Sorghastrum nutans* *
Japanese Blood Grass *Imperatas cylindrica 'Red Baron'*
Lambs' Ear *Stachys byzantina*
Lavender Cotton *Santolina chamacyparissus*
Mondo Grass *Ophiopogon japonicus*
Pampas Grass *Corderia selloana*
Ribbon Grass *Phalaris arundinaceae picta*
Side Oats Grama *Bouteloua curtipendula* *
Trumpet Vine *Camsis radicans* *
Virginia Creeper Vine *Parthenocissus quinquefolia* *
Wisteria Vine *Wisteria sp.*
Yellow Archangel *Lamiastrum galeobdolon* *

Blood Grass

Northeast Texas, Northern Louisiana, & Southern Arkansas

(Tyler, Longview, Shreveport-Bossier City, Texarkana)

**denotes native to the 4-state area*

Trees:

American Beech Fagus grandifolia *
Bald Cypress *Taxodium distichum* *
Cedar Elm *Ulmus crassifolia* *
Chinese Pistache *Pistachio chinesis*
Crape Myrtle *Lagerstroemia indica*
Cut-leafed Sumac *Rhus typhina* *
Deodora Cedar *Cedrus deodora*
Downy Serviceberry *Amelanchier arborea* *
Flowering Dogwood *Cornus florida* *
Fringe Tree *Chionanthus virginicus* *
Golden Raintree *Koelreuteria paniculata*
Leland Cypress *Cupressocyparis leylandii*
Loquat *Eriobotrya japonica*
Japnaese Maple *Acer palmatum*
Live Oak *Quercus virginiana* *
Longleaf Pine *Pinus palustris* *
Pecan *Carya illinoensis* *
Pin Oak *Quercus palustris* *
Redbud *Cersis canadensis* *
Red Maple *Acer rubrum* *
Smoke Tree *Cotinus coggygria* *
Southern Magnolia *Magnolia grandiflora* *
Stellata Magnolia *Magnolia stellata* *
Swamp White Oak *Quercus bicolor* *
Sweet Bay Magnolia *Magnolia virginiana* *
Tulip Popular *Liriodendron tulipifera* *
Water Oak *Quercus nigra* *
Weeping Yaupon Holly *Ilex vomitaria 'Pendula'* *

Shrubs:

Althea *Hibiscus syriacus*
American Holly *Ilex opaca* *
Camellia (Sasanqua) *Camellia sasanqua*
Carolina Buckthorn *Rahamnus caroliniana* *
Cherry Laurel *Prunus caroliniana* *
Cleyera *Ternstroemia gymnanthera*
Chinese Photinia *Photinia serrulata*
Evergreen Jasmine *Jasminum Sp.*
Farkleberry *Vaccinium arboreum* *
Frazer Photinia *Photinia x frazeri*
Gardenia dwarf *Gardenia jasminoides 'Radicans'*
Glossy Abelia *Abelia grandiflora*
Golden Groundsel *Senecio obovatus* *
Gold Dust Aucuba *Aucuba japonica 'variegata'*
Green Pittosporum *Pittosporum tobira*
Indian Hawthorne *Rhaphiolepis indica*
Japanese Boxwood *Buxus japonica*
Juniper *Juniperus sp.*
Lavender Tree (Chaste Bush) *Vitex agnus-castus* *
Leatherleaf Mahonia *Mahonia bealei*
Lily of the Nile *Agapanthus africanthus*
Nandina *Nandina domestica*
Oleander *Nerium oleander*
Pomegranate *Punica granatum*
Possumhaw *Ilex decidua* *
Possumhaw Viburnum *Viburnum nudum* *
Savannah Holly *Ilex x attenuata 'Savannah'* *
SkullCap *Scutellaria sp.* *

Longleaf Pine

Mayapple

Deodora Cedar

Nandina Domestica

Loquat

Wisteria

Snowball Bush *Viburnum sp.*
Stork's Bill *Erodium texanum* *
Virginia Sweetspire *Itea virginica* *
Waxleaf Ligustrum *Ligustrum japonica*
Wax Myrtle *Myrica cerifera* *
Weigela *Weigela florida*
Yaupon Holly (dwarf) *Ilex vomitoria 'nana'* *

Flowers and Herbs:

Cardinal Flower

Bearded Iris *Iris xiphiodes*
Big Blue Lobelia *Lobelia siphilitica*
Blue-Gray Sedge *Carex glauca* *
Canna Canna *x generalis*
Cardinal Flower *Lobelia cardinalis* *
Crinum Lily *Crinum americanum* *
Dame's Rocket *Hesperis matronalis*
Daylily *Hemerocallis sp.*
Drummond Phlox *Phlox drummondii* *
Four O'clock Mirabilis jalapa *
Gaura *Gaura lindheimeri* *
Gayfeather *Liatris pycnostachya* *
Hardy Hibiscus *Hibiscus Moscheutos* *
Indian Blanket *Guillardia sp.*
Lantana *Lantana sp.* *
Louisiana Iris *Iris fulva* *
Meadowrue *Thalictrum dasycarpum* *
Mealy Cup (Blue Sage) *Salvia farinacea* *
Mulleins *Verbascum sp.*
Pincushion Flower *Scabosa columbaria*
Red Hot Poker *Kniphofia uvaria*
Stonecrop *Sedum sp.* *
Sunray Coreopsis *Coreopsis grandiflora*
Sweet Violet *Viola odorata*
Thrift *Phlox paniculata*
White Avens *Geum canadense* *
Wormwood *Artemesia sp.*
Wood Sorrel *Oxalis crassipes*
Yarrow *Achillea millefolium* *

Spirea

Clematis Vine

Vines, Ornamental Grasses, and Groundcovers:

Ajuga *Ajuga reptans*
Asian Jasmine *Trachelospermum asiaticum*
Clematis Vine *Clematis sp.*
Crossvine *Bignonia capreolata* *
Coral Honeysuckle Vine *Lonicera sempervirens* *
Carolina Jasmine Vine *Gelsemium sempervirens* *
English Ivy *Hedera helix*
Creeping Fig Vine *Ficus pumila (repens)*
Fountain Grass *Pennistum*
Hosta *Hosta sp.*
Mayapple *Podophyllum peltatum* *
Mondo Grass *Ophiopogon japonicus*
Perwinkle *Vinca minor*
Carpet and Climbing Rose *Rosa sp.*
Trumpet Vine *Camsis radicans* *
Virginia Creeper Vine *Parthenocissus quinquefolia* *
Wisteria Vine *Wisteria sp.*

Camelia

Texas Panhandlen (Amarillo), Oklahoma Panhandle, and Northwest Oklahoma

**denotes native to the 4-state area*

Trees:

American Plum *Prunus americana* *
Arborvitae *Thuja orientalis*
Austrian Pine *Pinus nigra*
Black Locust *Robinia pseudoacacia* *
Colorado Blue Spruce *Picea pungens 'Glauca'*
Lacebark Elm *Ulmus parvifolia*
Chinese Pistache *Pistachia chinesis*
Hackberry *Celtis occidentalis* *
Honey Mesquite *Prosopis glandulosa* *
Kentucky Coffee Tree *Gymnocladus dioica* *
Nanking Cherry *Prunus tomentosa*
One Seed Juniper *Juniperus mono sperma* *
Pinon Pine *Pinus edulis* *
Ponderosa Pine *Pinus ponderosa* *
Purpleleaf Plum *Prunus cerasifera*
Rocky Mountain Juniper *Juniperus scopu.orum* *
Shatung Maple *Acer truncatum*
Western Soapberry *Sapindus drumondii* *

Purpleaf Plum

Ponderosa pine

Shrubs, Yucca, and Cactus:

Adams Needle Yucca *Yucca filmentosa* *
Eleagnus *Eleagnus fruitlandi*
Curveleaf Yucca *Yucca recurvifolia* *
Dwarf Burford Holly *Ilex comuta rotunda burfordii*
Flowering Quince *Chaenomeles japonica*
Lavender Cotton (Gray) *Santolina chamaecyparissus*
Junipers *Juniperus sp.*
Mugo Pine *Pinus mugo*
Oregon Grape Holly *Mahonia aquifolium*
Persian Lilac *Syringa persica*
Prickly Pear Cactus *Opuntia phaecantha* *
Rose *Rosa sp.*
Russian Sage *Perovskia atriplicifolia*
Soapweed Yucca *Yucca glauca* *
Sweet Olive *Osmanthus fragrans* *
Willowleaf Cotoneaster *Cotoneaster salicifolia*
Winterfat *Ceratoides lantana eurotia* *

Blackeyed Susan

Cotoneaster

Flowers, Grasses, and Vines:

Amelia's Sand Verbena *Abronia amelia* *
Bearded Iris *Iris xiphioides*
Bearded Tongue Penstemon *Penstemon gloxinioides* *
Blue Fescue Grass *Festuca ovina glauca*
Buffalo Grass *Buchloe dactyloides**
Hardy Aster *Aster novae-angliae*
Black-eyed Susan *Rudbekia hirta* *
Blue marguerite *Felicia amelloides*
Canna *Canna hybrids*
Clematis *Clematis sp.*
Columbine *Aquilegia* *
Daylily *Hemerocallis*
False Dragonhead *Physostegia virgiana* *
Feverfew *Chrysanthemum parthenium*
Foxglove *Digitalis*
Gloriosa Daisy *Rudbekia hirta*
Hollyhock *Althea rosea*
Light Poppy Mallow *Callirhoe alcaeoides* *
Phlox *Phlox paniculata*
Plains Melampodium *Melampodium leucanthum* *
Plains Zinnia *Zinnia grandiflora* *
Shasta Daisy *Chrysanthemum maximum*
Stonecrop *Sedum sp.* *
Violets *Viola odorata*

Silvermound

Extreme Southeast Texas

(Houston and Beaumont)

denotes native to the 4-state area

Trees:

Cherry laurel *Prunus caroliniana* *
Citrus *Citrus sp.*
Fringetree *Chionanthus virginicus* *
Laurel Oak *Quercus laurifolia* *
Live Oak *Quercus virginiana* *
Loquat *Eriobotrya japonica*
Red Bay *Perseaborbonia*
Slash Pine *Pinus elliottii* *
Southern Magnolia *Magnolia grandiflora*
Southern Red Oak *Quercus falcata* *
Strawberry Tree *Arbutus unedo*
Water Oak *Quercus phellos* *
Willow Oak *Quercus nigra* *
Winged Elm *Ulmus alata* *

Shrubs:

Acuba *Acuba japonica*
Azalea *Rhododendron sp.*
Bottlebrush *Callistemon lanceolatus*
Camellia *Camellia japonica*
Clera *Ternstroemia gymnanthera*
Coastal Leucothoe *Leucothoe axillaris* *
Fatshedera *Fatshedera lizei*
Fatsia *Fatsia japonica*
Feijoa *Feijoa sellowiana*
Fetter Bush *Lyonia lucida*
Florida Anise (Starbush) *Illicium floridanum**
Gardenia *Gardenia jasminoides*
Gold Star Esperanzaa *Tecoma stans 'Lonesp'*
Hibiscus *Hibiscus rosa-sinesis* *
Indian Hawthorn *Raphiolepsis sp.*
Ixora *Ixora coccinea*
Japanese Yew *Podocarpus macrophyllus*
Mexican Orange *Choisyaternata*
Nandina *Nandina domestica*
Oleander *Nerium oleander*
Pineapple Guava *Feijoa sellowiana*
Pittisporum *Pittisporum tobira*
Pomegranate *Punica granatum*
Sasanqua Camellia *Camellia sasanqua*
Strawberry Bush *Euonymous americana* *
Sweet Olive *Osmanthus fragrans* *
Sweet Viburnum *Vibrunum odoratissimum*
Texas Pistache *Pistacia texana* *
Texas Sage *Leucophyllum texanum* *
Asiatic Jasmine *Trachelospermum asiaticum 'variegata'*
Virginia Willow *Itea virginica* *
Wax Myrtle *Myricacerifera* *
Witch Hazel *Hamamelis virginiana* *
Yaupon *Ilex vomitoria* *

Palms:

Cabbage Palm *Sabal palmetto*
Canary Island Date Palm *Phoenix canariensis*
Dwarf Palmetto *Sabal minor* *
Louisiana Palmetto *Sabal Louisiana* *
Mediterranean Fan Palm *Chamaerops humilis*
Mexican Fan Palm *Washingtonia filifera* *
Needle Palm *Rhapidophyllum hystrix*
Pindo Palm *Butia capitata*

Pineapple Guava

Palmetto

Fatsia

Dwarf Palmetto

Azelea

Texas Palmetto *Sabal texana* *
Windmill Palm *Trachycarpus fortunei*

Ferns:
Christmas Fern *Polystichum acrostichiodes* *
Holly Fern *Cyrtomium falcatum*
Ebony Spleenwort *Asplenium platyneuron* *
Leatherleaf Fern *Rumohra adiantiformis*
Mariana Maiden Fern *Thelypteris torresiana*
Sword Fern *Polystichum munitum*
Variegated Shield *Archniodes simplicior*

Flowers:
Agapanthus (Blue Lily of the Nile) *Agapanthus africanus*
Amaryllis *Hippeastrum hybrids*
Bear's-Breeches *Acanthus spinosus*
Bog Lily *Crinum americanum* *
Bog Sage *Salvia Uliginosa* *
Brazilian verbena *Verbena bonariensis*
Canna *Canna sp.*
Chameleon Plant *Hymenocallis caroliniana*
Daylily *Hemerocallis*
Goldenrod *Solidago sp.* *
Jacobinia *Justicia carnea*
Louisiana Blue Phlox *Phlox divaricata* *
Louisiana Iris *Iris fulva* *
Marguerite *Chrysanthemum frutescens*
Mexican Marigold *Tagetes lucida* *
Mist Flower *Eupatorium*
Milkweed *Asclepias sp.* *
Nun's Orchid *Phaius grandifolius* *
Plumbago *Plumbago ariculata*
Purple Heart *Secreasea pallida 'Purple Heart'*
Shrimp Plant *Justicia brandegeana*
Society Garlic *Tulbaghia violacea*
Spider Lily *Hymenocallis*
Swamp Hibiscus *Hibiscus militaris* *
Swamp Lily *Hymenocallis caroliniana*
Texas Lantana *Lantana horrida*
Tuberose *Polianthes tuberosa*

Groundcover, Vines, and Grasses:
Algerian Ivy *Hedera canariensis*
Angel-wing Jasmine *Jasminum nitidum*
Asparagus Fern *Asparagus sprengeri*
Baby's Tears *Soleirolia soleirolii*
Chi Chi Ruellia *Ruellia 'Chi Chi'*
Coastal Muhly Grass *Muhlenbergia capillaris**
Cordgrass *Spartina patens* *
Confederate Jasmine *Trachelospermum jasminoides*
Carolina Jasmine Vine *Gelsemium sempervirens* *
Coral Vine *Antigonon leptopus*
Cross Vine *Bignonia capreolata* *
Everbloom Shrub Daisy *Euryops pectinatus 'Viridis'*
Evergreen Miscanthus Grass *Miscanthus transmorrisonensis*
Evergreen Wisteria Vine *Millettia reticulata*
Fall Ruby Grass *Rycheytrum neriglume*
Feather Reed Grass *Briza media*
Houston Meadowrue *Thalictrum texanum* *
Kiwi Fruit Vine *Actinidia chinesis*
Moneywort *Lysimachia nummularia*
Monkey Grass *Liriope muscari*
Pepper Vine *Ampelopsis arborea*
Prairie Cordgrass *Spartina pectinata* *
Swamp Jessamine Vine *Gelsemium rankinii*
Trailing Lantana *Lantana montevidensis* *
Wedelia *Wedelia tribolata*
Weeping Lovegrass *Eragrostis curvula* *

Blue Salvia

Hardy Hibiscus

Golden Moneywort

Lantana

Monkey Grass, 'Silverdragon'

Central Texas
(Austin, Waco, College Station, San Antonio)

**denotes native to the 4-state area*

Trees:

Aristocrat Pear Pyrus calleryana 'Aristocrat'
Bald Cypress Taxodium distichum *
Bigelow Oak Quercus sinulata breviloba *
Bigtooth Maple Acer grandidentatum *
Canaert Redcedar Juniperus virginiana *
Cedar Elm Ulmus crassifolia *
Chinquapin Oak Quercus muhlenbergii *
Chinese Pistache Pistachia chinesis
Crabapple Malus sp.
Desert Willow Chilopsis linearis *
Eldarica Pine Pinus eldarica
Escarpment Live Oak Quercus fusiformis *
Evergreen Sumac Rhus vierns *
Eve's Necklace Sophora affinus *
Fan d'Arc Osage Orange Maclura pomifera 'Fan d'arc'
Golden Raintree Koelreuteria paniculata
Hackberry Celtis laevigata *
Honey Mesquite Prosopis glandulosa *
Japanese Black Pine Pinus thunbergii
Japanese Persimmon Diospyros glandulosa
Lacey Oak Quercus glaucoides *
Live Oak Quercus virginiana *
Mexican Plum Prunus mexicana *
Mountain Cedar Juniiperus ashei *
Redbud Cercis sp. *
Shumard Oak Quercus shumardii *
Texas Mountain Laurel Sophora secundiflora *
Texas Red Oak Quercus texana *
Wafer Ash Ptelatrifoliata *
Western Soapberry Sapindus drummondii *

Chinquapin Oak Tree

Historic Live Oak Tree

Western Soapberry

Shrubs and Yuccas:

Abelia Abelia grandiflora
Agarito Mahonia trifolata *
Burford Holly (dwarf) Ilex cornuta rotunda burfordii
Barberry Berbis thunbergii atropurpurea
Cherry Laurel Prunus carolinia *
Chinese Photinia Photinia settulata
Eleagnus Eleagnus fruitlandi
Fern Acacia Acacia angustissima hirta *
Japanese Boxwood Buxus japonica
Junipers Juniperus sp.
Kidneywood Eysenhardtia texana *
Nandina (dwarf) Nandina sp. 'gulf stream'
Possumhaw Ilex decidua *
Pomegranate Punica granatum
Rusty Blackhaw Viburnum Viburnum rufidulum *
Texas Palmetto Sabla texana *
Texas Sotol Dasylirion texanum *
Torrey Yucca Yucca torreyi
Twistleaf Yucca Yucca rupicola *
Turk's cap Malvaviscus drummondii *
Variegated Pittosporum Pittosporum tobira variegata
Woolybucket Bumelia Bumelia lanuginosa *

Nandina Nana

Evening Primrose

Flowers:

Autumn Sage Salvia greggi *
Baby Blue Eyes Nemophylla phaceliodes
Blackeyed Susan Rudbekia x goldstrum *
Bluebells Eustoma grandiflora
Blue Plumbago Plumbago auriculata
Bluebonnets Lupinus texensis *
Butterfly Weed Asclepias tuberosa *
Canna Canna x generalis
Chilipequin Caspicum annuum *
Cigar Plant Cuphea micropetala
Clasping Coneflower Dracopsis amplexicaulis *
Damianita Chrysactinia mexicana *
Daylily Hemerocallis sp.
False Dayflower Commelinantia anomala *
Firebush Hamelia patens
Fountain Grass Pennisetum ruppelii *
Frogfruit Phyla nodiflora *
Gayfeather Leatris sp. *
Goldeneye Viguiera dentata *
Golden Groundsel Senecio obovatus *
Indian Blanket Gaillardia sp. *
Jerusalem Artichoke Helianthus tuberosus *
Lantana Lantana sp. *
Lemon Mint Monarda citriodora *
Mealy Cup Sage Salvia leucantha *
Mexican Hat Ratibda columnifera *
Mock Bishopweed Bifora americana *
Moonbeam Coreopsis Coreopsis verticillatan *
Obedient Plant Phystogia virginiana *
Pink Evening Primrose Oenothera sp. *
Purple Coneflower Echinacea purpurea *
Rainlillies Zephyranthes grandiflora
Rosemary Rosmarinus officinalis
Verbena Verbena sp.
Velvet Senna Senna lindheimeriana *
Wild Foxglove Penstemon cobaea *
Yarrow Achillea millefolium *
Zemenia Wedelia hispida *

Vines, Grasses, and Groundcover:

Ajuga Ajuga reptans
Asiatic Jasmine Trachelosperum asiaticum
Boston Ivy Parthenocissus tricuspidata
Carpet Rose and Climber Rosa sp.
Cedar Sedge Carix planostachys *
Confederate Jasmine Trachelosperum jasminoides
Coral Honeysuckle Lonicera sempiverens *
Fig Ivy Ficus Pumila repens
English Ivy Hedera helix
Grape Vine Vitis sp.
Muhly Grass Muhlenbergia lindheimeri *
Pampas Grass Cortaderia selloana
Santolina Santolina sp.
Silverlace Vine Polygonum aubertii
Switchgrass Panicum virgatum *
Virginia Creeper Parthenocissus quinquefolia *
Wisteria Wisteria sp.

Moonbeam Coreopsis

Coneflowers

Yarrow

Boston Ivy

North Central Texas

(Dallas, Fort Worth, Denton, Wichita Falls)

** denotes native to the 4-state area*

Trees:

Bald Cypress *Taxodium distichum* *
Burr Oak *Quercus macrocarpa* *
Cedar Elm *Ulmus crassifolia* *
Chinese Pistache *Pistachio chinesis*
Chinkapin Oak *Quercus muhlenbergia* *
Crabapple *Mallus sp.*
Crape Myrtle *Lagerstroemia indica*
Deodora Cedar *Cedrus deodora*
Desert Willow *Chilopsis linearis* *
Canert Redcedar *Juniperus virginiana* *
Eldarica Pine *Pinus eldarica*
Flameleaf Sumac *Rhus lanceolata* *
Honey Mesquite *Prospis glandulosa* *
Japanese Black Pine *Pinus thunbergiana*
Live Oak *Quercus virginiana* *
Mexican Plum *Prunus mexicana* *
Pecan *Carya illinoensis* *
Red Buckeye *Aesculus pavia* *
Redbud *Cersis canadensis* *
Shumard Oak *Quercus shumardii* *
Smoke Tree *Cotinus coggygria* *
Southern Magnolia *Magnolia grandiflora 'Little Gem'* *
Texas Ash *Fraxinus texensis* *
Texas Madrone *Arbutus texana* *
Virginia Pine *Pinus virginiana*
Yaupon Holly *Ilex vomitaria* *

Leatherleaf Mahonia

Redbud and Whitebuds

Shrubs:

Agarita *Berbis trifoliolata* *
Althea *Hibiscus syriacus*
Carolina Buckthorn *Rahamnus caroliniana* *
Chinese Holly *Ilex cornuta*
Clasping Coneflower *Rudbekia amplexicaulis**
Crape myrtle *Lagerstroemia indica*
Eleagnus *Eleagnus fruitlandi*
Fragrant Mimosa *Mimosa borealis* *
Frazer Photinia *Photinia x frazeri*
Glossy Abelia *Abelia grandiflora*
Golden Groundsel *Senecio obovatus* *
Gold Dust Aucuba *Aucuba japonica 'variegata'*
Green Pittosporum *Pittosporum tobira*
Indian Hawthorne *Rhaphiolepis indica*
Japanese Boxwood *Buxus japonica*
Juniper *Juniperus sp.*
Lavender Tree (Chaste Bush) *Vitex agnus-castus* *
Leatherleaf Mahonia *Mahonia bealei*
Nandina *Nandina domestica*
Oleander *Nerium oleander*
Pomegranate *Punica granatum*
Rusty Blackhaw *Viburnum rufidulum* *
Possum haw *Ilex decidua* *
SkullCap *Scutellaria sp.* *
Snowball bush *Viburnum sp.*
Stork's Bill *Erodium texanum* *
Texas Sage (Cenizo) *Leucophyllum frutescens**
Virginia Sweetspire *Itea virginica* *

Weigela

Gold Dust Acuba

Weigela *Weigela florida*
Yaupon Holly (dwarf) *Ilex vomitoria 'nana'* *

Flowers and Herbs:

Autumn Sage *Salvia greggii* *
Bearded Iris *Iris xiphiodes*
Blue-Gray Sedge *Carex glauca* *
Canna *Canna x generalis*
Crinum Lily *Crinum sp.*
Damianita *Chrysactinia mexicana*
Four O'clock *Mirabilis jalapa*
Garden Canna *Canna x generalis*
Gaura *Gaura lindheimeri* *
Grape Hycinth *Muscari thinbergenianum*
Hardy Hibiscus *Hibiscus moscheutos* *
Hartweg (Yellow) Evening Primrose *Calylophus hartwegii* *
Indian Blanket *Guillardia sp.* *
Lanceleaf Coreopsis *Coreopsis lanceolata**
Lantana *Lantana sp.* *
Meadowrue *Thalictrum dasycarpum* *
Mexican Petunia *Ruellia sp.* *
Ox-eye Daisy *Chrysanthemum leucanthemum**
Pincushion Flower *Scabosa columbaria*
Prarie Verbena *Verbena bipinnatifida* *
Red Hot Poker *Kniphofia uvaria*
Rockrose *Pavonia lasiopetala* *
Shadscale *Atriplex confertifolia* *
Snow-on-the-Prairie *Euphorbia bicolor* *
Sunray Coreopsis *Coreopsis grandiflora*
Sweet Violet *Viola odorata*
Sweet Viburnum *Vibrunum odoratissimum*
Texas Bluebell *Eustoma grandiflorum* *
Texas Star Daisy *Lindheimera texana* *
White Avens *Geum canadense* *
Winecup *Callirhoe involucrata* *
Wormwood *Artemesia sp.*
Wood Sorrel *Oxalis crassipes*
Yarrow *Achillea millefolium* *

Vines, Ornamental Grasses, and Groundcovers:

Asian Jasmine *Trachelospermum asiaticum*
Carolina Snailseed Vine *Cocculus carolinus*
Cast Iron Plant *Aspadistra elatior*
Crossvine *Bignonia capreolata* *
Coral Honeysuckle Vine *Lonicera sempervirens* *
Carolina Jasmine Vine *Gelsemium sempervirens* *
English Ivy *Hedera helix*
Creeping Fig Vine *Ficus pumila (repens)*
Fountain Grass *Pennistum sp.*
Hosta *Hosta sp.*
Mondo Grass *Ophiopogon japonicus*
Muhly Grass *Muhlenbergia lindheimeri*
Pampas Grass *Corderia selloana*
Passion Vine *Passiflora incarnata* *
Carpet and Climbing Rose *Rosa sp.*
Red Barrenwort *Epumedium rubrum*
Side Oats Grama *Bouteloua curtipendula* *
Switch Grass *Panicum virgatum* *
Strawberry Begonia *Saxifraga sarmentosa*
Trumpet Vine *Campsis radicans* *
Virginia Creeper Vine *Parthenocissus quinquefolia* *
Wisteria Vine *Wisteria sp.*
Yellow Archangel *Lamiastrum galeobdolon*

Candy Tuft

Texas Star Daisy

Red Barrenwort

Yellow Archangel

South Texas

(Corpus Christi, Del Rio, and Brownsville)

**denotes native to the 4-state area*

Trees, Yuccas & Palms:

Anaqua *Ehretia anacua*
Beaked Yucca *Yucca rostrata* *
Brasil (Bluewood) *Condalia hookeri* *
California Fan Palm *Washingtonia filifera*
Florida Fan Palm *Sabal palmetto*
Guajillo Acacia *Acacia berlandieri* *
Live Oak *Quercus virginiana* *
Honey Mesquite *Prospis glandulosa* *
Mexican Fan Palm *Washingtonia robusta*
Red Bay *Persea borbonia* *
Roemer Acacia *Acacia roemeriana* *
Sweet Acacia *Acacia smallii farnesiana* *
Texas Ebony *Pithecellobium flexicaule*
Texas Mountain Laurel *Sophora secundiflora* *
Texas Sabal Palm *Sabal texana*
Trecul Yucca *Yucca treculeana*
Wax Myrtle *Myrica cerifera* *
Wild Olive *Corida boissieri* *
Windmill Palm *Trachycarpus fortunei*

Mexican Fan Palm

Cast Iron Plant

Shrubs:

Agarito *Mahonia trifoliata* *
Cape Honeysuckle *Tecomaria capensis*
Cast Iron Plant *Aspidistra elatior*
Catsclaw Mimosa *Mimosa biuncifera* *
Cenizo (Texas Sage) *Leucophylla frutescens* *
Century Plant *Agava americana* *
Coral Bean *Erythrina herbacea* *
Esperanza *Tecomastans angustata*
Flame Acanthus *Anisacanthus quadrifidus*
Firecracker *Russelia equisetiformis*
Firebush *Hamelia patens*
Firespike *Odontonemastrictum*
Oleander *Nerium oleander*
Pink Lantana *Lantana camara* *
Sago Palm *Cycas revoluta*
Skeleton Leaf Goldeneye *Viguiera stenolba* *
Turk's Cap 'drummondi' *Malvaviscus arboreus* *

Sago Palm

Flowers, Vines and Groundcover:

Autumn Sage Salvia greggii *
Bougainvillea *Bougainvillea sp.*
Coralvine *Antigononleptopus*
Butterfly Weed *Asclepias curassavica* *
Copper Canyon Daisy *Tagetes lemmonnii*
Dogweed *Dyssodia pentachaeta* *
Firebush *Hamelia patens*
Gayfeather *Liatris montevedensis* *
Heartleaf Hibiscus *Hibiscus cardiophylus* *
Ice plant *Carprobrotus edulis*
Lazy Daisy *Aphanostephus skirrhobasis*
Lemon Mint *Monarda Citriodora* *
Mexican Flame Vine *Senecio confusus*
Mexican Sage *Salvia leucantha* *
Mist Flower *Eupatorium coelestinum*
Natal Plum *Carissa grandiflora* *
Pampas Grass *Cortaderia selloan*
Passion Flower Vine *Passiflora incarnata* *
Penta *Penta lanceolata*
Prostrate Rosemary *Rosemary officinallis*
Trumpet Vine *Campsis radieans* *
Verbena Verbena *sp.* *
Weeping Lovegrass *Eragrostis curvula*
Winter Honeysuckle *Lonicera fragrantissima**

Bouganvillea

West Texas

(El Paso, Lubbock, Abilene, Midland/Odessa, San Angelo, Abilene)

** denotes native to the 4-state area*

Trees:

Arizona Cypress *Cupressus arizonica* *
Autumn Amber Sumac *Rhus aromatica flabelliformis* *
Chisos Red Oak *Quercus gravesii* *
Chiltapa Tree *Chilopsis linearis x Catalpa biginoniodes*
Desert Willow *Chilopsis linearis* *
Eldarica Pine *Pinus brutia eldarica*
Honey Locust *Gleditsia triacanthos*
Honey Mesquite *Prosopis galandulosa* *
Lacey Oak *Quercus laceyi (glaucoides)* *
Littleleaf Ash *Fraxinus Greggii*
Mexican Buckeye *Ungnadia speciosa* *
Shinny Xylosma *Xylosma senticosa*
Shrubby Dogwood *Thymophylla acerosa* *
Texas Madrone *Arbustus texensis* *
Texas Mountain Laurel (Mescal Bean) *Sophora secundiflora* *
Western Soapberry *Sapindus drummondii* *

Arizona Blue Cypress

Shrubs: Yucca, and Cactus:

Apache Plume *Fallugia paradoxa* *
Barberry *Berbis thunbergii*
Beebush *Aloysia grassima* *
Black Dalea *Dalea frutescens* *
Bottlebrush *Callistemon lanceolatus*
Brewer Saltbush *Atriplex lentiformis 'Breweri'*
Candelilla *Euphorbia antisyphilitica* *
Chihuahuan Sage *Leucophyllum laevigatum*
Creosote Bush *Larrea tridentata* *
Dwarf Coyote Brush *Baccharis pilulasis*
Indian Hawthorne *Raphiolepsis indica*
Juniperus *Juniperus sp.*
Lavendar Cotton *Santolina chamaecyparissus*
Leuchugullia *Agave lechuguilla* *
Ocotillo *Fouqueria splendens*
Pineapple Guava *Feijoa sellowiana*
Red Yucca *Herperaloe parviflora* *
Shrubby Blue Rock Sage *Salvia pinguifolia* *
Shrubby Boneset *Eupatorium havanense*
Shrubby Senna *Cassia wislizeni*
Soaptree Yucca *Yucca faxoniana*
Sugar Bush *Rhus ovata*
Rubber Rabbit Bush *Chrysothamnus nauseosus* *
Variegated Euonymous *Euonymous 'Aureo-variegata'*
Willow Leaf Cotoneaster *Cotoneaster salicifolia*
Yaupon Holly (dwarf) *Ilex vomitoria nana* *
Yellow Bird of Paradise *Caesalpinia gilliesii*
Yellow Trumpet Flower (Yellow Bells) Tecoma stans *

Yarrow

Agave

Flowers:

Beardlip Penstemon *Penstemon barbatus* *
Damianita *Chrysactinia mexicana* *
Desert Marigold *Baileya multiradiata* *
Mexican Gold Poppy *Eschsholzia californica 'mexicana'* *
Mexican Hat *Ratibda* *
Lantana *Lantana sp.* *
Parralena (Dogweed) *Thymophylla pentachaeta* *
Parry Penstemon *Penstemon parryi* *
Pink Plains Penstemon *Penstemon ambiguus**
Rock Penstemon *Penstemon baccharifolius**
Sand Aster *Aster canescens* *
Scented Lippia *Lippia gravolens* *
Snapdragon Vine *Maurandya antirrhiniflora* *
Thorn Apple *Datura wrightii* *
Verbena *Verbena sp.* *
Yellow Bells *Tecoma stans* *
Yellow Daisy *Rudbekia hirta* *

The Green Goods Industry

Plants that are sold in nurseries and garden centers are produced from a variety of sources. Generally, smaller container-grown plants (1 to 5 gallons) are grown by large wholesale growers who have a vast array of greenhouses and areas for growing. Plants grown there have the benefits of almost total climate control. They are protected from pests and funguses, floods and droughts, extreme high and low temperatures. Unfortunately, plants don't necessarily end up for sale in the zones where they do best. Air-conditioned trucks transport plants to retail stores where the new environment may be dramatically different from where the plants were originally grown.

Plants are sold three different ways. Plants sold in category one are grown in plastic gallon containers and wooden boxes. In category two, plants are balled and wrapped in burlap or grown in root control bags. Root control bags are fabric bags that wrap around the roots and adjacent soil of the small trees. Plants are grown in fields; when they are ready for sale, the owner digs up the root control bags, risking less damage to the growing tree's root system. In category three, the plants (trees) are grown in fields and later dug up bare rooted and then sold.

Container grown plants provide a year round planting opportunity. If the ground is not frozen and plants can be watered after planting, then container grown plants can be planted. The best time to plant (except for cold sensitive evergreens) is in the fall, followed by winter, spring, and then summer.

Planting Category One Plants

Be sure to inspect your prospective plant before you make a purchase. If the stem appears wobbly or loose and not firmly a part of the root ball, make another selection.

Look for well-balanced plants; in other words, if the plant appears disproportionately large in comparison to the container, then it has outgrown its optimal container size.

Color is an indicator of plant health and vigor. Avoid those plants that have burnt or yellowing leaves and patchy foliage.

Check the root system of plants once they are removed from their container. If they are "root bound" (have circling roots), the roots will have a hard time growing into the surrounding soil.

To prevent difficulties, take a razor or knife and slash three or four vertical cuts across the circling roots.

Dig a hole 1 1/2 to 2 times the width of the container. Backfill with soil from the site.

Container grown Colorado Blue Spruce

After planting, a mulch layer of 3 to 4 inches can be applied.

Planting Category Two Plants

Balled and burlapped, and root control bagged trees are usually the most expensive plants due to their height and trunk diameter (caliper inch) size. Evergreens are usually priced by the foot. Deciduous trees are priced by the caliper inch. Slower growing shade trees such as oaks normally cost more per inch than faster growing trees such as an ash or elm. Plants are marketed during the spring after the fall or winter harvest. Nurseries typically keep ball and burlapped plants in a bed of mulch in which they are watered and cared for.

Dormant trees -- means this is the time of year to harvest them.

To plant a B and B tree, start by digging a hole that will be 1 1/2 to 2 times the width of the ball being planted.

Set the ball on the undisturbed soil at the bottom of the dug out hole.

The top of the ball should be at the same height the tree was originally growing in the field. Roots usually grow out horizontally or at a slight angle to the surface where oxygen, nutrients and water are generally more favorable.

Once it is in the hole, and if you're sure the tree is facing from its best side, remove any twine that is tied around the trunk. Failure to remove non-rotting twine results in the girdling and eventual death of a tree.

Peel back the burlap that covers the top of the root ball. Then either cut and remove the burlap or pull it down the side of the root ball without disturbing the ball and back fill.

When planting a tree grown in a root control bag, follow the same procedure as above for the B and B stock.

Cut 4 vertical slashes through the fabric and peel back the root control fabric, much as you would peel a banana.

Before back filling, remove the plastic under the bottom of the ball.

Be sure that all fabric has been removed before you back fill.

Planting Category 3 Plants

Bare rooted plants are normally sold during winter (January and February). Plant bare root plants before they have leafed out. Fruit and nut trees, roses, ornamental shrubs, grapevines, and blackberries are often sold bare rooted. They are generally packed with a moist medium such as sawdust or sphagnum moss.

Before planting, trim any broken or dead roots, and soak for up to 12 hours or as directed.

The most important thing about planting bare rooted stock is that the roots should never be allowed to dry out.

The hole should be 1 1/2 to 2 times the width of the root system.

Place the plant so that it is at the same level as it was growing in a field.

Splay roots in the hole so that they lay out naturally.

Greenhouse flowers

Back fill and press firmly around the roots so as to prevent any air pockets.

Make a dam around the plant and soak well.

Basic Plant Requirements

Soil

Before you start planning your garden, take a good look at the soil you have. Test the soil, and make any soil adjustments if necessary. See below for treatment suggestions.

There are 10 orders of soil in the 4-state region. You'll need to dig at least six to eight inches deep for an accurate soil sample. Texas Agriculture Extension Service, Louisiana Cooperative Extension Service, The Arkansas Cooperative Extension Service, and the Oklahoma State Cooperative Extension Service all have soil testing facilities. Also, nurseries and garden centers sell reliable soil testing kits that can test for pH, nitrogen, phosphorus, and potassium. (for web site or phone numbers of the above state agencies see the listings at the bottom of page 30.)

Acidic soils tend to occur in areas of heavy rainfall (the Gulf Coast, most of Arkansas and Louisiana, southeast Oklahoma, and areas of east Texas). See the rainfall map. Rainfall leaches naturally occurring alkali chemical elements from the soil. Alkaline soils occur in the arid regions of the high plains and much of the prairie. If you are interested in the general types of soil for your area, contact the Soil Conservation Service. They have maps that show the different soil types, usually on a county by county basis.

Soil scientists have identified over 75,000 kinds of soil in the United States. An average soil sample is 45% minerals, 25% water, 25% air, and 5% organic matter.

Below are three samples that show different soil texture. The proportions of sand, silt, and clay are what give soil texture. Which one of the samples has the best soil texture?

Sand

Sandy soil has coarse particles. This provides good drainage and airspace. Sand does not retain water for long periods and nutrients leech easily. Additions of organic matter help the soil retain water and nutrients longer.

Loam

Clay, sand, and silt are all in balanced proportions. The clay portion helps to retain water and nutrients. Sand allows enough air space for roots to grow and to provide drainage. Soil that is high in organic matter normally has brown to blackish color, the color of our sample.

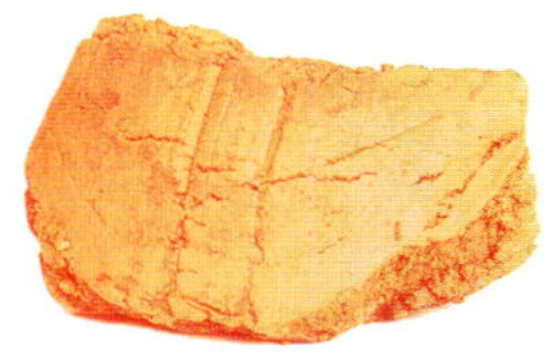

Clay

Clay has the smallest mineral particles of the samples. Clay soil leaves little room for airspace. This inhibits the ability of roots to penetrate soil. On a plus side, clay retains moisture and nutrients. An addition of organic matter and or sand allows roots to better penetrate.

*Note the smooth glaze of the clay sample. The glaze further impedes root penetration Adding coarse sand or compost will counteract the glazing effect.

Treatments

When a soil pH is too low, you would add limestone because this will raise the pH. If you want to lower the pH on a mildly alkaline soil (7.5 - 8.5), you would add sulfur or mulch such as peat moss, pecan hulls, or pine needles. Consult your state agriculture agent or local nurseryman on the amounts necessary to make corrections.

Is your soil more clay-like or sandy? A simple test reveals the answer. See illustration. Clay soils have tiny pore spaces between small clay particles, so they don't drain well. The lack of air space makes it more difficult for roots to take up nutrients, and they can rot. Sandy soils have relatively large particles, drain well, and offer good air exchange. Shortcomings include the tendency to drain quickly, the relative inability to retain nutrients, and the tendency for plants to wilt quickly.

A clay soil needs the increased ability to hold air and disperse water. In order to improve use, you should add generous amounts of coarse sand or organic matter like compost, finely ground wood mulch, or humus to break up the small clay particles. Additions of compost, wood mulch, straw, or other organic matter will help sandy soil retain water and nutrients.

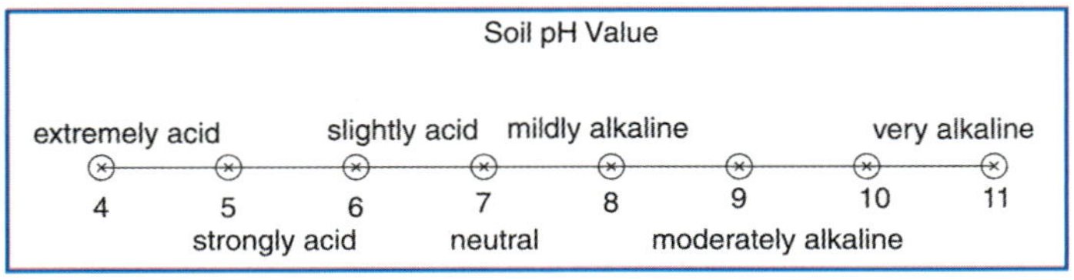

pH Readings

pH is the chemical reading of how well a soil is balanced. The pH scale runs from 0 (extremely acid) to 14 (extremely alkaline). 0-7 is the acid range; 6.5 is mildly acid and is desirable for acid-loving plants; 7-14 is the alkaline range. pH levels tell you how well the plants can use the nutrients in the soil. Nitrogen, phosphorus, and potassium are the three primary nutrients essential for plant growth. Nitrogen, the single most important element, is responsible for the aboveground growth of the plants. Phosphorus contributes to the formation of seeds, fruit and root growth. Potassium increases the resistance of some plants to disease. In low pH soil (under 6), the necessary nutrients aren't available, so elements such as aluminum and molybdenum (which are in abundance) become toxic. In high pH soil (higher than 8), magnesium would be more readily available. Just as a low pH impacts certain nutrients adversely, so does a high pH impact other nutrients adversely. Iron, manganese, boron, copper, and zinc (trace minerals) are less available in an alkaline soil and more available in an acid soil. Also, either extreme of pH is detrimental because of the effects on bacterial and fungal activity that breaks down organic matter.

Water

You might think water is readily available for plant intake, since three quarters of the earth's surface is covered by either liquid water or frozen water. However, the percentage of fresh water is 3% compared to 97% that constitutes the world's oceans.

Water is the single most important substance into which nutrients and essential elements are dissolved, transported, and eliminated in plants. As water is deposited as rain or irrigation, it travels through the soil dissolving materials in its path. These nutrients and materials are then absorbed by plant roots for use in many chemical reactions within the plant. Trees are made up of up to 80% water. Plants require such large amounts of water in order for photosynthesis to work. In addition, water is the major component for heating and cooling plants.

We call the mechanism for cooling plants: evapo-transpiration. This is the combination of 1) evaporation of water and water vapor from the surface of plant tissues into the surrounding atmosphere, and 2) the release of water vapor during respiration of the plant. As plants respire, they "breathe" out water vapor and waste gases generated from chemical processes by cells within the plant. As moisture is lost in the surrounding atmosphere, this creates an imbalance in the concentration of water at the top of the plant when compared to the concentration of water in the root zone. Water then moves up the plant toward the areas of the plant that have lost water.

Some people think trees have a "tap root" growing deep into the soil that allows them to escape drought. This is not the case. While some species of trees do develop tap roots, they are mainly for stabilizing the tree and preventing the tree from falling over in windy conditions. Actually, the small roots and fine hairs occupying the top 2-2.5 feet of soil provide the nutrient and water uptake for trees. The tremendous amount of these small feeder roots and root hairs provides an enormous amount of surface area exposed to both soil nutrients and soil moisture.

"How much water should I apply to my plants, and how often should I water them?"

For large established trees and shrubs, water should penetrate at least 18 inches into the soil. Using a "soaker hose" is an ideal way to deliver irrigation slowly so that water is transferred down through the soil profile, exposing many of the feeder roots and root hairs to the much needed moisture. As with your lawn, it is best to water long, slow and deep as opposed to many frequent shallow waterings. After you have thoroughly wet the soil profile containing the "feeder" roots, you can judge when you need to water again by digging down in to the soil at the dripline (this is where the branches end) of the tree. If you have to dig down deeper than two inches to find moist soil, it is time to turn the "soaker hose" on again. This is the best method to judge dryness because the amount of water used by plants varies according to the amount of sunlight, wind speed, and temperature the plants are exposed to.

If your trees or shrubs are new plantings, water once per week. During their second year, water every two weeks, and during their third year, every three weeks. After the third year, adjust water schedules depending upon the amount of natural rainfall received.

Smaller shrubs should be watered with a "soaker hose" so that water moves 12 inches into the soil. After that, monitor the soil moisture by digging around the edge of the shrub with your finger. If you have to dig down deeper than one inch to find moist soil, it's time to water again. Keep in mind that these are general recommendations, and some plants require more water than others.

While water is an important part of maintaining your plants' health, remember it is only one factor involved in the complex workings of trees, shrubs, and other ornamentals. Poor plant health is usually the result of multiple stress factors at work at the same time. If you understand how these stresses affect plant growth, you'll be able to remedy them before your plants suffer.

Extension Listings

University of Arkansas Cooperative Extension Service
501.671.2000, fax 501.671.2209 www.arhomeandgasrden.org

Louisiana State University Cooperative Extension Serrvice
225.578.2065, fax 225.578.2526 www.agctrlsu.edu

Oklahoma State University Cooperative Extension Service
405.744.5385, fax 405.744.5339 www.agweb.okstate.edu/pearl

Texas Agricultural Extension Service
979.845.7800, fax 979.845.9542 www.agextension.tamu.

Designs

The key to a successful garden or landscape project is a plan or landscape design. Without an overall plan, selecting plants for your garden can be a hit or miss proposition. What is the purpose of the area you've designated to be your garden? Is it to be a restful spot where you can relax and enjoy the flowers? Are you trying to attract nature's creatures: birds, butterflies, rabbits, to your garden? Or are you trying to divert attention from some unpleasant aspects of your property? Whatever the purpose, remember that form follows function. Your overall design for your garden area will reflect the purpose of the space. Three important variables: style, structure, and color, will help you achieve your purpose.

Additionally, you will need to consider:

How will my garden change as the plants grow and the seasons change?

How much maintenance will be required on my part to keep this area healthy and thriving?

If I don't have much money to spend all at once, can I develop a garden in stages?

Have I considered the soil requirements for my planned garden, and am I making the best use of sunlight and shade?

How will my garden look at different times of the day?

Remember, unlike a building, a garden is constantly growing and changing.

Style

You want your garden to reflect your personality style, which ought to blend with the surrounding architecture as well. What do I mean by style? Does a formal style with close-clipped and manicured hedges in a straight line fit with your personality? Or do you feel more comfortable with a natural style that attempts to mimic a nearby ecosystem? Does the formal style seem too stiff and high maintenance for you? Does the natural style seem too unkempt and unruly to you? Look at the way you live, and consider if you're more formal or more free-spirited.

Natural Look

Formal Garden

Structure

When you plan your garden space, look at ways to incorporate existing structures (trees, pathways, brooks, and rocks) into the design. Structure is anything from a patio to a pathway, trellis to fence, major trees and other focal points. Rather than ignore or tear down existing structures, use them in your design scheme.

Color

You can create different effects for your garden by carefully choosing the color of your flowers and plants. For example, blues, pinks, and purples are cool and calming colors. They also give the illusion of a shallow space being deeper. If you had only a small garden, and wanted the space to appear larger, you would want to use blues and purples.

On the other hand, bright yellow and reds are warm colors and convey vibrancy. You want to plant warm colors up close to brighten an area. Using variegated foliage, reds, or burgundies spark more interest than their simple green counterparts. These brighter colors will attract more attention to your space, but will overwhelm your audience if used too lavishly.

Sometimes you might use bright colors to divert attention from a different area in your yard. Bright colors sprinkled in the middle of a garden with whites and pastels will draw people's eyes away from unsightly corners and angles.

The goal in choosing colors is to harmonize. Colors opposite each other on the color wheel should be used sparingly as accents. Using different shades of green plants and grasses ties the garden together, as green is the basic color of any garden.

Unfortunately, you also need to know something about your plants, not just the color of their flowers and leaves. If you're dealing with annuals, you can more easily predict the color scheme. If you're using perennials, you have to consider their different blooming times in order to blend and harmonize their colors. However, strategically placing certain colors together is pleasing to the senses and will help you achieve your overall purpose of respite or drama for your garden.

Color Wheel of Plants

The color wheel is a combination of red, blue, and yellow, the three primary colors. All of the other colors in between the primary colors are combinations of the mixing of the three colors. Notice each plant that is associated with its color. Combinations of colors, which are neighbors on the color wheel, have a harmonizing effect. The use of colors opposite on the color wheel has a contrasting or dramatic effect.

If you are unsure how certain plants will look together, then try cutting out pictures from old plant catalogs and patch them together. See which grouping suits you the best. This method saves a lot of time and energy especially if you are planning a large garden.

Nursery and Garden Centers

Arkansas

See page 68 for survey details.

501 Area Code - Ft. Smith, Little Rock, Fayetteville And Area

A Touch Of Color
8207 Rogers Ave Fort Smith, AR 72709 501-484-1020

A.W. Shrubs
Route 2 Box 114B Lonoke, AR 72086 501-843-3731

Alpine Nursery
Route 1 Box 14 Altus, AR 72821 501-468-3611

American Bonsai Nursery
5141 Industrial Park Rd. Van Buren, AR 72956 501-474-9225
Del; Bus.9 yrs; Nat; Bonsai with all species; Exp; Demo G;
Quarterly-E-mail, hours by appointment.

Amos' OrchaRd. & Nursery
18292 Wildlife Rd. Fayetteville, AR 72701 501-443-0282

Arkansas Garden Center
15603 Cantrell Rd. Little Rock, AR 72212 501-868-0103

Arkansas Gardens
9511 Highway 107 Sherwood, AR 72120 501-834-4018

Arkansas Plant Brokers
9325 Ferndale Cut-Off Little Rock, AR 72223 501-868-8733

Arkanwood Nursery
30603 Hwy 300 Rolan, AR 72135 501-330-2893

Bamis Tree Farm, Inc.
13206 Asher Rd. Little Rock, AR 72206 501-897-4931

Barbara's Gardens
2503 Hiram Davis Place Fayetteville, AR 72703 501-442-0905

Barbara's Gardens
2148 North Thompson Springdale, AR 72764 501-750-2818

Nursery Specialties Key

Del - Delivery
Bus- Years in operation
% plant stock grown w/in 75 mile radius
Nat - Native plants
Exp - 2 years minimum staff experience
Mem - Member of nursery association
Learn C - Learning center
Fert - Organic fertilizer & pesticide
Demo G - Demonstration garden
News - Newsletter
T Spade - Tree spade

See page 68 for survey details.

Battelfield Nursery
136 Hwy 62 Avoca, AR 72711 501-925-3556

Baucum Nursery -AR Forestry Commission
1402 Hwy 391 North N. Little Rock, AR 72117 501-945-3345

Blackburn Nursery
111 Locust Conway, AR 72032 501-329-2143

Blossom Nursery
216 CR 326 Eureka Springs, AR 72632 501-253-7895
Bus. 3 yrs; 33%; Nat; Paw paws, persimmons other natives; Exp 2 yrs;
Open by appointment, and mail order; www.blossomnursery.com

Blossomberry Nursery
Rural Route5, Box 28-F Clarksville, AR 72830 501-754-6489
Del; Bus 30yrs; 33%; Nat; Unusual plants, Fruit trees, Land/Design; Exp;
Mem; Fert; Hours 8-5 M-S, Sun 2-5.

B-Mart Garden Center
110 East Illinois St. Beebe, AR 72012 501-882-3307

Bradford Nursery
1605 Prairie Creek Dr. Rogers, AR 72757 501-636-8393

Bradford Nursery
Hwy E. 12 Rogers, AR 72756 501-636-8393

Breckling Nursery
1122 Gold Links Rd. Hot Springs, AR 71901 501-624-5847

Breshears Nursery
109 Brookside Trace Hot Springs, AR 71909 501-624-6208

Bur Oak Farm
15959 Hamstring Rd. Fayetteville, AR 72703 501-582-3617

Burks' Nursery
10 Hyton St. Benton, AR 72015 501-315-6453

Burnett's Landscape Supply & Nursery
10501 Maumelle Blvd N. Little Rock, AR 72113 501-771-2456

Burnett's Landscape Supply & Nursery
9709 Warden Rd. N. Little Rock, AR 72113 501-833-2033

Burnetts Landscape Supply & Nursery
9709 Warden Rd. Fayetteville, AR 72703 501-582-3617
Del; 33%; Nat; Bonsai; Exp; Mem; Learn C; Fert

Cabot Lawn & Garden Center, Inc.
1001 Linda Lane Cabot, AR 72023 501-843-5523

Cantrell Gardens, Inc.
7800 Cantrell Rd. Little Rock, AR 72207 501-225-1030

Carden-Harris Azalea Nursery
Route 1 Box 53 Branch, AR 72928 501-965-2972

Cenark Growers
2293 South Market St. Benton, AR 72015 501-315-7592

Cleburne County Nursery
1901 West Quitman St. Heber Springs, AR 72543 501-362-8522

Conway Gardens And Pools
2740 Prince Conway, AR 72032 501-329-1358

Country Market Garden C&R
P.O. Box 576 Bald Knob, AR 72010 501-724-3134

Cricket Hill Farms
45 Sunny Gap Rd. Conway, AR 72032 501-327-7520

D's Discount Nursery
#1 Brannan Rd. Greenbrier, AR 72058 501-679-6017

Dan's Nurseries, Inc.
415 Hwy. 11 Searcy, AR 72143 501-742-3830

Davidson & Co. & Conway Nursery, Inc.
571 Hwy 64 E. Conway, AR 72032 501-327-2362

Davidson Nursery
922 Access Rd. Beebe, AR 72012 501-882-5530

Davis Tree Farm
241 Crook Pangbura, AR 501-728-4036

Didier's Garden Center
P.O. Box 6512 Ft. Smith, AR 72916 501-646-6222

Dobb's Mountain Nursery
22029 Dobbs Mountain Rd. Evansville, AR 72729 501-848-3609
Del; Bus. 6 yrs.; Trees, Paperbark Maples, Hostas & Perennials; Exp; Demo G; Learn. C; Fert; Call for hours.

Doc's Greenhouse & Garden
5635 E Huntsville Rd. Fayetteville Ark. 72701 501-442-2061

Drew Wood Nursery
35 Stiers Lane Cabot, AR 72023 501-941-5717

Emerald Green Growers Inc.
3334 Fair Park Blvd. Little Rock, AR 72204 501-565-1937

Ever-Green Lawn & Nursery & Landscaping Co.
1709 N. Thompson Springdale, AR 72764 501-751-6760

Nursery Specialties Key

Del - Delivery
Bus- Years in operation
% plant stock grown w/in 75 mile radius
Nat - Native plants
Exp - 2 years minimum staff experience
Mem - Member of nursery association
Learn C - Learning center
Fert - Organic fertilizer & pesticide
Demo G - Demonstration garden
News - Newsletter
T Spade - Tree spade

See page 68 for survey details.

Evergreen Nursery
P.O. Box 7 Centerville, AR 72829 501-576-2794

Family Garden Nursery
P.O. Box 147 Pettigrew, AR 72752 501-677-2308
Del; Bus.- 17 yrs.; 33%; Exp; Mem; Fert; Hours M-F 8:30 - 3:00

Four Seasons Produce & Garden Center
1104 W. Pleasure St. Searcy, AR 72143 501-268-4966

Freddy's Nursery
601 Pittman Rd. Hot Springs, AR 71913 501-767-4054

Freddy's Nursery
1838 Airport Rd. Hot Springs, AR 71913 501-767-4054

Fryer's Tree Farm
33332 SW 14th St. Bentonville, AR 72712 501-795-2778
Bus. 28 yrs; 33%; Nat; Exp; Hours 8:30-5 Tu-F, 8-4 Sat, closed S-M.

Ft. Smith Garden Center
4100 Planters Rd. Ft. Smith, AR 72908 501-646-0100

Galbraith Greenhouse & Nursery
HC 72 Box 10 Clinton, AR 72031 501-745-4204

Galbraith Greenhouse & Nursery
No. Sycamore Dr. Mt. View, AR 72533 501-745-4204

Garden City Nursery
2105 S. Walton Blvd. Fayetteville, AR 72712 501-273-5928

Garden City Nursery
1610 E. Tahlequah St. Siloam Springs, AR 72761 501-524-6961

Garden Of Eden Nursery & Landscape
3520 Central Ave. Hot Springs, AR 71913 501-623-6616

Garzini's Gardens & Gifts
1260 N. Main St. Waldron, AR 72958 501-623-6616

Good Earth Garden Center & Nursery, Inc.
15601 Cantrell Rd. Little Rock, AR 72223 501-868-4666

Green Country Nursery & Landscaping, Inc.
P.O. Box 362/6200 Wilkerson St. Johnson, AR 72741 501-521-5500

Groen Nursery
3414 S. Hwy 252 Charleston, AR 72953 501-965-7834
Del; Bus 12yrs; Crape Myrtles; Demo G; Fert.

Green Thumb Garden Center Of Little Rock, Inc.
12000 Kanis Rd. Little Rock, AR 72211 501-227-5454

Green Tree Nursery & Landscape Co. Inc.
9305 Rodney Parham Rd. Little Rock, AR 72227 501-225-6313

Harmony Ridge Farm
403 Shilow Rd. McRae, AR 72102 501-882-0944

Harris Greenhouse
P.O. Box 396 Dardanelle, AR 72834 501-229-3304

Highlander Blueberry Nursery
P.O. Box 177 Pettigrew, AR 72752 501-677-2300
Del; Bus 15yrs; 33%; Blueberries; Exp; Mem; Fert; Open by appt.

Hill's Greenhouse & Nursery
250 Copperas Gap Rd. Houston, AR 72070 501-889-5260

Hocotts Garden Center, Inc.
3612 Kavanaugh blvd. Little Rock, AR 72205 501-663-8376
Bus.22 yrs; Demo G; Exp; Mem; Fert; Hours M-Sat 8-5, Sun 12:30 - 5

Horticare Wholesale Nursery, Inc.
7901 Stagecoach Rd. Little Rock, AR 72214 501-407-2727

Hosta Hill
16 Towering Oaks Dr. Conway, AR 72032 501-329-1940

Huber Nursery
123 Hwy 321 Beebe, AR 72012 501-882-6498

Hurlbut Nursery & Landscaping
285 Elm Springs Rd. Elm Springs, AR 72728 501-248-1130

Jackson's Greenhouse & Landscaping
4001 West Main Russellville, AR 72801 501-968-2101

Kay's Nursery
10203 Jacksonville-Cato N. Little Rock,AR 72120 501-835-3948

Kountry Market
11804 Sweetwater Run N. Little Rock, AR 72117 501-945-9576

Lafoon's Greenhouse And Nursery
8934 Rogers Ave. Fort Smith, AR 72903 501-452-4494

Lakewood Gardens
3101 North Hills Blvd. N. Little Rock, AR 72116 501-753-7800
Del; Bus. 16yrs; 33%; Nat; Perennials; Mem; Learn C; Fert;
Hours 8-5:30 M-Sat, Sun 12-5; www.lakewoodgardens.com

Lathrop's Nursery & Landscapes, Inc.
601 W. Martin Heber Springs, AR 72543 501-362-3029

Leo W. Lafoon Jr. Greenhouses & Nursery
8934 Rogers Ave. Ft. Smith, AR 72903 501-452-1636

Living Garden Nursery
534 N. Front St. Fisher, AR 72429 501-238-7689

Logan Nursery
3117 Warwick Fayetteville, AR 72703 501-872-0369
15638 Osage Hill Rd. Siloam Springs, AR 72761 501-443-7154
Del; Bus.4yrs; 33%; Exp; Mem.

M&J Plants
2710 Sheryl Ave. Fayetteville, AR 72703 501-444-7480

Maumelle Nursery & Landscape
11700 Maumelle Blvd. N. Little Rock, AR 72113 501-758-0002

Meumeier Nursery & Florist, Inc.
3327 North O St. Ft. Smith, AR 72904 501-782-9987

Morningside Garden & Nursery, Inc.
1170 Hwy 64 West Morrilton, AR 72110 501-354-8470

Moss Rock Nursery & Landscape Co.
3026 S Hwy 161 N. Little Rock, AR 72117 501-982-1079

My Secret Garden
P.O. Box 37 Lowell, AR 72745 501-925-1485

New Life Nursery & Landscaping
1400 Antioch Rd. Donaldson, AR 71941 501-865-8478

Old Country Greenhouse
14987 South Hwy 59 Siloam Springs, AR 72761 501-549-3698

Ouachita Highland Farm
246 Ouachita Highland Rd. Pencil Bluff, AR 71965 501-326-4482

Ouachita Nursery
37913 Hwy. 67 Donaldson, AR 71941 501-384-2283

Ozark Bamboo Garden
1059 Cr 266 Eureka Springs, AR 72631 501-253-6801

Ozark Perennials
11584 Jordan Lane Prairie Grove, AR 72753 501-846-2719

P.A.C. Plant Farm
P.O. Box 1029 Judsonia, AR 72081 501-729-3326

Pangburn Nursery
117 Stringfellow Rd. Pangburn, AR 72121 501-728-3502

Paw Paw Products
216 CR 326 Eureka Springs, AR 72632 501-253-7895
Bus 10yrs; 33%; Nat; Native Plants; Exp; www.blossomnursery.com

Perennials, Etc.
15905 Hwy 62 Garfield, AR 72732 501-359-3412

Pine Ridge Gardens
832 Sycamore Rd. London, AR 72847 501-293-4359
Lmtd Del; Bus 9 yrs; Nat; Over 300 species of native plants,Bird & Butterfly natives; Mem; Demo G; Open by appt; www.pineridgegardens.com

Plants Plus
845 Longview Fayetteville, AR 72703 501-521-1445

Plants Plus Nursery & Garden Center
845 Longview Fayetteville, AR 72703 501-521-1445

Pottsville Nursery
5801 Hwy 64E Pottsville, AR 72858 501-968-7849

R/C Nursery
3671 Hwy 8 W. Mena, AR 71953 501-394-4519

River Valley Horticultural Products & Services
21701 Lawson Rd. Little Rock, AR 72210 501-821-4770

Rogers Greenhouse
1402 Buttercup Lane Paragould, AR 72450 501-239-9496

Simmons Plant Farm
11542 N. Hwy 71 Mountainsburg, AR 72946 501-369-2345

Southaven Nursery & Collectables
Rt. 3, Box 234A Ozark, AR 72949 501-667-2405

Spurlin Plant Farm
1577 Shady Grove Rd. Hot Springs, AR 71901 501-262-1930

Strawberry Fields Nursery, Inc.
1112 N. Maple St. Searcy, AR 72143 501-268-0725

The Plant Doctor & Sons Nursery
9109 Zero St. Ft. Smith, AR 72903 501-452-6631

The Tree Transplanters
3806 E Central | Bentonville, AR 72712 | 501-751-4484

Three D Nursery
390 Hwy 11 | Searcy, AR 72143 | 501-742-3867

TLC Nursery & Supply
1701 Bracy Rd. | Little Rock, AR 72206 | 501-888-1444

Unique Crafts, Greenhouse, Counseling Clinic
307 E. Sevier St. | Benton, AR 72015 | 501-776-0297

Westwood Gardens
4324 Weddington Dr. | Fayetteville, AR 72703 | 501-442-3929

Westwood Gardens
2104 E. Mission Blvd. | Fayetteville, AR 72703 | 501-571-1500

Whitehaven Farms
37 Reservoir Heights Dr. | Little Rock, AR 72227 | 501-227-7674

Willow Springs Christmas Tree Farm
13724 Sandy Ann Drive | Little Rock, AR 72206 | 501-888-1129

870 Area Code - Pine Bluff, El Dorado, and Jonesboro

Abundant Daylilies & Perennials
HC 78 Box 187 | Star City, AR 71667 | 870-628-4820

Al Williams & Son Nursery, Inc.
P.O. Box 155 | Piggott, AR 72454 | 870-598-3357

Azalea Hill Gardens & Nursery
1106 S. Evans Rd. | Pine Bluff, AR 71602 | 870-247-1574

Backyard Nursery & Landscaping
308 Raney St. | Hoxie, AR 72433 | 870-886-1755

Bed O' Roses
980 Lakehall Rd. | Lake Village, AR 71653 | 870-265-4418

Bennett's Nursery & Garden Center
110 Southwest Dr. | Jonesboro, AR 72401 | 870-932-3180

Black Forestry Ltd.
2018 S. Main St. | Stuttgart, AR 72160 | 870-673-2242

Bloomin' Baskets
P.O. Box 1447 | Glenwood, AR 71943 | 870-356-2522

Broadway Nursery-Landscaping
2023 Hwy 62 Box 3 | Pocahontas, AR 72455 | 870-892-9341

Nursery Specialties Key

Del - Delivery
Bus- Years in operation
% plant stock grown w/in 75 mile radius
Nat - Native plants
Exp - 2 years minimum staff experience
Mem - Member of nursery association
Learn C - Learning center
Fert - Organic fertilizer & pesticide
Demo G - Demonstration garden
News - Newsletter
T Spade - Tree spade
See page 68 for survey details.

Brown's Nursery
136 Binnis Loop | Crossett, AR 71635 | 870-364-6779

Carl Wayne's Nurseries, Inc.
1151 East 9th | Mountain Home, AR 72653 | 870-425-6966

Carroll F. Quarry
299 Lawrene 532 | Alicia, AR 72410 | 870-886-3091

Chapel Gardens Nursery
5000 Chapel Garden Dr. | Pine Bluff, AR 71603 | 870-879-1070

Cosby's Greenhouse & Garden Center
1412 Parkway Dr. | Crossett, AR 71635 | 870-364-2130

Country Road Nursery
Rte. 2 Box 154 | Hampton, AR 71744 | 870-798-4973

Dick Gonce Nursery
703 East 9th St. | Mountain Home, AR 72653 | 870-425-7722

Ellis' Greenhouse & Nursery
3988 Hwy 133 North | Hamburg, AR 71646 | 870-853-8156

Four Seasons
111 Hamilton | Camden, AR 71701 | 870-836-4444

Four Seasons Nursery & Florist
3484 W. State Hwy 18 | Manila , AR 72442 | 870-561-3570

Grannies Bloomers
107 E. 7th | Ashdown, AR | 870-898-8515

Green Thumb Nursery
6900 South Olive | Pine Bluff, AR 71603 | 870-535-4707

Hideaway Garden & Landscaping
Rt. 1 Box 192d | Des Arc, AR 72040 | 870-256-4173

Hodges Nursery
4431 Hwy 1 | Cherry Valley, AR 72324 | 870-588-4444

Holman's Azalea Garden
P.O. Box 54 | Wickes, AR 71973 | 870-385-2262

Hope Nursery & Garden Center, Inc.
P.O. Box 27 | Hope, AR 71802 | 870-777-8834

Jerry's Nursery & Landscaping, Inc.
619 Main St. | Arkadelphia, AR 71923 | 870-230-1678

Kates Acres Greenhouse
5366 Stadium Blvd. | Jonesboro, AR 72401 | 870-935-1965

Ladybug Nursery
540 Ewen St. | Weiner, AR 72479 | 870-684-2325

LakeSue Nursery
1410 Columbia 36 | Magnolia, AR 71753 | 870-234-4477

Larry Byers Nursery & Greenhouse
3918 Highway 29 S. | Hope, AR 71801 | 870-777-3543

Legacy Herbs
HC 70, Box 442 | Mountain View, AR 72560 | 870-269-4051

Living Spring Nursery
1060 Clearview Rd. Pocahonats, AR 72455 870-892-8580

Main Street Gardens
717 Hwy 83 S. Monticello, AR 71655 870-382-2255

Merlo Gardens
2505 E. 10th Ave. Pine Bluff, AR 71601 870-534-4943

Mid South Nursery
P.O. Box 15106 Jonesboro, AR 72403 870-932-2036

Moore's Garden Center & Nursery
104 Dover Rd. West Memphis, AR 72301 870-732-3729

Old Time Flower Gardens
County Rd. 17 Genoa, AR 870-773-5082
Perennials

Ozark Springs Nursery
304 Bomber Blvd. Mountain Home, AR 72653 870-425-3800

Ozark Wildflower Co.
HC 70 Box 169 Jasper. AR 72641 870-446-5629
Over 250 different perennials; open by appt; www.pineridgegardens.com

Paradox Water Gardens
39 Mc 3046 Yellville, AR 72687 870-427-8688

Pittman Garden Center
1305 W. 19th El Dorado, AR 71730 870-862-8733

Pittman Nursery Corp.
700 Columbia Rd. 61 Magnolia, AR 71754 870-234-1111

Pocahontas Garden Center
387 Country Club Rd. Pocahontas, AR 72455 870-892-3041

Pruett's Nursery
1002 Malcolm Ave. Newport, AR 72112 870-523-3514

Ramsey Nursery
1311 Hwy 62/65 N. Harrison, AR 72601 870-741-8808
Del; Bus.18yrs; Perennials, Annuals, Herbs, Container gardens;
Demo G; Exp; Mem; Learn C; Fert; Hours M-Sat 8-5:30,
call for Sun hours; e-mail - ramsey@alltel.net

Ridgecrest Nursery -N- Gardens
3347 Hwy 64 Wynne, AR 72396 870-238-3763
Del; Bus 16yrs; 33%; Nat; Native; Demo G;
email: Ridgecrest@crosscountybank.com

Ron's Garden Center
106 W. Hillsboro El Dorado, AR 71730 870-863-8385

Rosebud Nursery
3409-Center Ridge Rd. Amity, AR 71921 870-342-5334

Rosewood Farm
HC 71 Box 595 Mountain View, AR 72560 870-269-9298
Bus 3yrs; 33%; Nat; Exp; Demo G; Learn C; Fert; Hours 9-5 T-Sat.

Ross Nursery
1501 N. College St. El Dorado, AR 71730 870-863-8149

Shepard Florist & Greenhouses, Inc.
910 West 29th Ave. Pine Bluff, AR 71603 870-535-0911

Sieberts Nursery & Landscaping
1306 Hwy 49 North Paragould, AR 72450 870-236-1995

Son-Shine Nursery
2531 Hwy 26 Highland Hardy, AR 72542 870-856-3291

Southlawn Nursery, Inc.
3703 S. Culberhouse Rd. Jonesboro, AR 72404 870-931-9024
Del; Bus- 5yrs; Nat; Exp; Mem; News; T-Spade; Hours M-F 8-6, Sat 8-5.

Spider's Web Daylily Garden
2926 Hwy 150 Blytheville, AR 72315 870-763-5512

Spring Creek Nursery
1036 Fendley Rd. Amity, AR 71921 870-342-5839

That Bloomin Place
216 South Moore Rd. Hot Springs, AR 71913 801-767-9875

The Orchid Lady
20 Winnebago Trace Cherokee Village, AR 72529 870-257-2625

The Plant Outlet
#3 Chapel Village Pine Bluff, AR 71603 870-879-2570

The Wright Place
1229 Hwy 133 N. Crossett, AR 71635 870-364-3123

903 Area Code- Texarkana

Twitty Nursery, Inc.
4815 W. 7th St. Texarkana, AR 75501 903-838-5591

Louisianna

225 Area Code - Baton Rouge Area

Batton Rogue Horticultural Service
2222 Bardwell Dr. Baton Rogue, LA 70806 225 -924-5884

Bodin's Nursery
2580 Hwy. 20 West Vacherie, LA 70090 225-265-1080

Burris's Nursery And Greenhouses
8318 Mickens Rd. Baton Rogue, LA 70811 225- 356-3356

Callais Nursery
58030 Robertson St. Plaquemine, LA 70764 225-687-4272

Clegg's Nursery Inc.
274 N Donmoor Av. Baton, Rogue, LA 70806 225 927-1419

Dupont Nursery
58409-Belleview Dr. Plaquemine, LA 70764 225-687-8242

Garden District Nursery
2504 Government St. Baton Rogue, LA 70806 225 383-3514

Goodwood Hardware Nursery & Repair
1575 Lobdell Baton Rouge, LA 70806 225-926-0040

Gulf Coast Plantsmen
15680 Perkins Rd. Baton Rogue, LA 70810 225-751-0395

Hart's Garden Center
12655 S. Harrell's Ferry Baton Rouge, LA 70816 225-293-4922

Highland Hardware & Garden Center
7461 Highland Rd. Baton Rouge, LA 70804 225-766-3050

Hill's Nursery, LLC
Hwy 190 Livonia, LA 70755 225-637-3596

Juliette's Gardens LLC
6983 Jefferson Hwy Baton Rogue, LA 70806 225 -923-0516

Nature's Own Garden Center
15450 Terrell Rd. Baton Rogue, LA 70816 225 752-6997
Del; Bus 7yrs; 33%; Nat; Native plants; Exp; Fert; hrs. 9-6 M-S, Sun 12-5.

Pineywoods Nursery
31625 Courtney Cir. Walker, LA 70885 225-664-0473

Plant World
10131 Greenwell Springs Rd. Baton Rouge, LA 70814 225-272-7144

Plantech Nursery
14605 Airline Hwy Baton Rogue, LA 70817 225 753-1765

Pleasant Places, Clio Nursery
17535 Bull Run Rd. Springfield, LA 70462 225-695-3792

Pumpkin Center Plants & Pets
42018 Pumpkin Center Rd. Hammond, LA 70403 225-294-7387

Rose Hill Garden
4955 Hwy 955 Ethel, LA 70730 225-634-3355
Roses; open Feb-May 9-5; www.rosehillgarden.com

Randy's Nursery
1010 Etta Dr. St. Garbriel, LA 70776 225-642-8453

318 Area Code - Shreveport, Bossier City, Lafayette, Monroe, Lake Charles

ABC Orchids & Tropicals
3154 N E Evangeline Thrwy Lafayette, LA 70507 318-237-4795

All Seasons Nursery
2974 Johnston St. Lafayette, LA 70503 318-264-1418

Angel Gardens
403 Joli Rd. Lafayette, LA 70520 318-896-5399

Antique Rose Ville
2007 Freyou Rd. New Iberia, LA 70560 318-367-3000

Nursery Specialties Key

Del - Delivery
Bus- Years in operation
% plant stock grown w/in 75 mile radius
Nat - Native plants
Exp - 2 years minimum staff experience
Mem - Member of nursery association
Learn C - Learning center
Fert - Organic fertilizer & pesticide
Demo G - Demonstration garden
News - Newsletter
T Spade - Tree spade

See page 68 for survey details.

B&S Nursery
1244 Musser Rd. Lake Charles, LA 70611 318-855-3300

Barn, Yard, & Garden
105 First St. Jonesville, LA 71343 318-339-6695

Bayou Concrete Gardens
600 Jane St. New Iberia, LA 70563 318-367-5665

Belle's Nursery
La Hwy. 31 Arnaudville, LA 70512 318-754-5258

Betty's Plants & Shrubs
16086 Hwy 4 Chatham, LA 71226 318-249-2706

Blessings Of Nature Nursery
13760 Hwy 4 West Columbia, LA 71418 318-649-5277

Blue Bayou Nursery & Garden Center
1904 Moss St. Lafayette, LA 70501 318-269-0017

Bonsai Of Lafayette, Inc.
124 Darteze Dr. Lafayette, LA 70508 318-856-8491

Bud & Blooms Nursery
670 Pleasant Hill Cem. Rd. Plain Dealing, LA 71064 318-326-5726

C&K Gardens
10691 Crossett Rd. Bastrop, LA 71220 318-283-0477

Cathy's Plants & Herbs
1909 Eraste Landry Rd. Lafayette, LA 70506 318-989-9664

Chamberlain's Landscape & Nursery Inc.
1025 Central Park Dr. Bossier City, LA 318-742-5911

Choice Lawn Landscape & Nursery Center
425 Verot School Rd. Lafayette, LA 70508 318-237-7555

Cole's Sligo Road Garden Center
659 Sligo Rd. Bossier City, LA 71112 318-742-8889

Colletti's Plants
823 E. Main Jeanerette, LA 70544 318-276-5038

Country Gardens
158 Johnny Lane Leesville, LA 71446 318-239-2928

Creative Crafts Garden Center
124 Plank Rd. St. Joseph, LA 71366 318-766-8080

D&L Nursery & Greenhouse
344 Hwy 8 Lena, LA 71447 318-793-2832

Dee's Nursery
806-Fragala St. Rayville, LA 71269 318-728-5515

Donovan's Roses
150 Leo Ave. Shreveport, LA 71133 318-861-6693

Family Farm & Garden
780 W. Georgia Ave. Many, LA 71449 318-256-3167

Family Feed & Garden Center, Inc.
9333 Walker Rd. Shreveport, LA 71113 318-688-6001

Fleur-De-Lys Gardens & Accessories
433 Farmington Dr. Lafayette, LA 70503 318-981-0281

Foret & Foret Horticulturists
258 Bendel Rd. Lafayette, LA 70503 318-235-4366

Garden Of Eden
1670 Jack Crowell Rd. Eros, LA 71238 318-249-2729

Garrisons Greenwood Gardens
6905 Greenwood Rd. Shreveport, LA 71119 318-635-8296
Del; Bus 41yrs; Nat; Roses, shrubs & annuals; Exp; Mem; Demo G; Fert; Open 7 days a week 8:30-5:30

Girouard's Greenhouse & Nursery
309 South St. Jane Broussard, LA 70518 318-837-6859

Greenland Nursery
1835 Grand Anse Hwy Breaux Bridge, LA 70517 318-667-6436
Del; Bus 10 yrs; 33%; Nat; Exp; Learn C; Fert; Spring hours W-S 9-4, Sun 10-2, M & Tu closed.

Haughton Farm & Garden
102 N. Myrtle Haughton, LA 71037 318-949-9243

Hill Nursery
Hwy 15 Archibald, LA 71216 318-248-3141

Hilltop Farm & Garden, Inc.
860 Hwy 80 East Bossier City, LA 71006 318-949-1126

Hynum Nursery
#3 Crothers Dr. Talulah, LA 71282 318-574-5772

Jayroe's Hardware & Garden
207 5th St. Benton, LA 71006 318-965-0956

Kent's Nursery
Hwy 89 Youngsville, LA 70592 318-856-6716

Kinder Garden Center & Flower Shop
445 9th St. Kinder, LA 70648 318-738-5704

Kitchen's Green Thumb
12345 Crossett Rd. Bastrop, LA 71220 318-281-6991

Krotz Springs Home & Garden
24392 Hwy 190 Krotz Springs, LA 70750 318-566-8665

Lake View Nursery
Hwy 65 North Lake Providence, LA 71254 318-559-6044

Laviolette's Plant & Flower Center
110 St. Nicholas Dr. Lafayette, LA 70506 318-981-6904

Lex Plant Farm
9045 E. Kings Hwy Shreveport, LA 71115 318-797-6035

Logger Nursery
6551 Hwy 34 Chatham, LA 71226 318-249-4821

Louisana Nursery
Route 7 Opelousas, LA 70570 318-948-3636
Hostas

Marion Andrus Nursery
1924 Dulles Dr. Lafayette, LA 70506 318-981-3104

Marshall's Home & Garden Showplace
3550 Ambassador Caffery Pky Lafayette, LA 70503 318-984-3230

Marshall's Nursery Inc.
418 Rue De Belier Lafayette, LA 70506 318-984-3443

Midsouth Nursery
5495 Hwy 4 Ringgold, LA 71068 318-894-7301

Mr. Rogers Nursery & Landscaping
3601 Southern Ave. Shreveport, LA 71104 318-226-4010

My Father's Garden Gate Inc.
5675 Bert Kouns Industrial Lp. Shreveport, LA 71129 318-905-8045

Natural Garden & Nursery
1515 Johnson St. Jennings, LA 70546 318-824-3356

Neptune's Water Garden
2302 Washington St. Monroe, LA 71201 318-322-4801

Nia's Nursery
713-Louisiana St. Lake Arthur, LA 70549 318-774-2569

Nona's Plants
Hwy 3104 Midway Jena, LA 71342 318-992-2030

Orchid Gardens
1007 Breaux Bridge Ave. Lafayette, LA 70501 318-234-6339

Paradise Plantscapes
108 Lesin Broussard Rd. Lafayette, LA 70507 318-261-1125

Parterre
813 Park Ave. Monroe, LA 71201 318-323-7009

Plantland Greenhouse & Nursery
9998 Cross Creek Vivian, LA 71082 318-375-3726

Pommier's Plants, Etc.
5 Miles South On Hwy 330 Abbeville, LA 70510 318-898-2023

Porter's Brushy Creek Farm & Nursery
Hwy 84 E Winnfield, LA 71483 318-628-3123
Del; 10 Yrs; 33%; Nat; Perennials, shrub & trees; Exp 2 yrs; Open 8:30-6:00 daily

Pottery Gardens & Gifts
3816 W Congress St. Lafayette, LA 70506 318-989-9589

R&B House Of Plants
702 Hwy 109 South Starks, LA 70661 318-743-5413

Rainbow Gardens
4540 Hwy 80 Rayville, LA 71269 318-728-5429

Rainbow Nursery
108 Violet St. West Monroe, LA 71291 318-322-4167

Reliable Nursery
4710 Barksdale Blvd. Bossier City, LA 71112 318-741-1299

Robinson's Nursery
8222 Mansfield Rd. Shreveport, LA 71108 318-688-1168

Ross Gwinn Landscape & Nursery
1736 Washington Natchitoches, LA 71457 318-352-6166

S&S Greenhouse Nursery
1401 Hospital Ave. Franklin, LA 70538 318-828-1405

Schultz Nursery Enterprises
8300 A. Derouen Rd. Bell City, LA 70630 318-622-3436

Sears Garden Center
3601 Southern Ave. Shreveport, LA 71104 318-226-4010

Sharp Nursery & Landscape
Hwy 80 - Next To Serios Feed & Seed Bossier City, LA 71111 318-747-6400

Specialty Daylilies
124 Joel Sharp Rd. Farmerville, LA 71241 318-368-9420

Stagg's Nursery
106 E. Harrop Rayne, LA 70578 318-334-7740

Swartz Garden Market
1860 Hwy 139 Monroe, LA 71203 318-345-1080

TG & Associates Inc. Nursery Division
456 Hwy. 531 Minden, LA 71055 318-377-3750

Teche Nursery
3326 Verot School Rd. Youngsville, LA 70508 318-856-5908

The Flower Bed
Ridge Road @ Ambassador Caffery Lafayette, LA 70555 318-893-7449

The Gardens & Nursery Gift Shop
800 E Broussard Rd. Lafayette, LA 70508 318-981-6514

The Herb Cart
144 Emerite St. Lafayette, LA 70506 318-988-0663

Vicki's Nursery
100 E. TX St. Leesville, LA 71446 318-529-5775

Victoria's Garden Spot
122 Laura St. Abbeville, LA 70510 318-898-9484

Vinton Nursery Inc.
4912 Gray Rd. Vinton, LA 70668 318-589-7416

Water Oak Farms
1904 Hwy 178 Opelousas, LA 70570 318-942-7680

337 Area Code - Lafayette Area

Acadiana Nursery
2517 Avery Island Road New Iberia, LA 70560 337-364-2311

Nursery Specialties Key

Del - Delivery
Bus- Years in operation
% plant stock grown w/in 75 mile radius
Nat - Native plants
Exp - 2 years minimum staff experience
Mem - Member of nursery association
Learn C - Learning center
Fert - Organic fertilizer & pesticide
Demo G - Demonstration garden
News - Newsletter
T Spade - Tree spade

See page 68 for survey details.

Anthony's Nursery And Landscaping
3336 N. Pinhook Rd. Lafayette, LA 70508 337-237-8858

Blue Springs Nursery
167 Andre Martin Blvd. Opelousas, LA 70570 337-948-6571

Byrd's Nursery & Landscape Co.
7540 Johnston St. Lafayette, LA 70555 337-988-0323

D&R Plants Garden Center
5391 Frontage Rd. Opelousas, LA 70570 337-594-9606

Frank's Nursery
112 Demarest Rd. Grand Lake, LA 70607 337-598-3234

Lastrape's Garden Center, Inc.
5150 I-49 South Opelousas, LA 70570 337-942-3008

Main Street Garden
524 W. Main St. New Iberia, LA 70560 337-364-0987

Mr. Mcgregor's Garden
5770 Johnson St. Lafayette, LA 70503 337-981-8438

Pecan Creek Farms, LLC.
672 Zick Miller Rd. Sunset, LA 70584 337-662-6065

Rock Bottom Stone Co.
1550 Service Rd. I-49 N Grand Coteau, LA 70541 337-662-1100

504 Area Code - New Orleans Area

Abita Gardens Nursery
Lowe Davis Rd. & I-59 Abita Springs, LA 70420 504-796-5443

Alvarez Nursery
46188 N. Morrison Blvd. Hammond, LA 70401 504-345-4401

American Aquatic Gardens
621elysian Fields Av. New Orleans, LA 70117 504-944-0410

August J. Weber Garden Center
5007 Feret New Orleans, LA 70117 504-895-3022

Bamboo Gardens Of Louisiana
38124 Highway 440 New Orleans, LA 70450 504-795-2300

Banting Nursery
3425 River Rd. Bridge City, LA 70094 504-436-4343

Barker's Hdwe. & Nursery
14117 E. Main St. Cut Off, LA 70345 504-693-4332

Bennie's Garden Center
2301 Wash. St. Franklinton, LA 70438 504-839-5923

Bloomers Nursery & Garden
12845 Plank Rd. Baker, LA 70714 504-774-6263

Bordelon's Nursery
35813 La Hwy 16 Watson, LA 70726 504-665-2975

Brier Rose Herbs, Inc.
8632 Hiway 23 Belle Chasse, LA 70037 504-392-7499

Chackbay Nursery
837 Hwy 20 N. Thibodaux, LA 70301 504-633-9608

Charvet's Garden Center
4511 Clearview Pkwy Metairie, LA 70006 504-888-7700

De Battista's Nursery Inc.
210 Huntlee Dr. New Orleans, LA 70131 504-366-2017

Design With Nature
3835 River Road New Orleans, LA 70121 504-837-4564

Double Oak Garden Center
120 Hwy 654 Matthews, LA 70375 504-532-5160

Dubois Nursery
1188 W. Tunnel Blvd. Houma, LA 70360 504-851-4093

El's Nursery
178 Flanacher Rd. Zachary, LA 70791 504-654-3540

Faye's Garden Center
585 Holmes Blvd. Gretna, LA 70056 504-368-1322

First Street Nursery
2300 Tchopatoulous St. New Orleans, LA 70130 504-522-6022

Friends Of Hilltop Arboretum, Inc.
11855 Highland Rd. Baton Rouge, LA 70810 504-767-6916

Garden Spot Of Franklinton
1023 Wiliams St. Franklinton, LA 70438 504-839-9357

Garden Spot Of Mandeville
2100 North Causway Mandeville, LA 70470 504-674-0252

Green Acres Nursery, Inc.
731 Cardinal Dr. Thibodaux, LA 70301 504-447-6771

Greenway Nursery
24647 La Hwy 1032 Denham Springs, LA 70726 504-665-9925

Harb's Oasis, Landscaping & Garden Center
13827 Coursey Blvd. Baton Rouge, LA 70817 504-756-2720

Hometown Farm & Garden
301 Nw Central Ave. Amite, LA 70422504-748-9347

Inwood Gardens
1640 Collins Blvd. Covington, LA 70433 504-892-0717

Island Nursery Farm
61287 Vining Rd. Amite, LA 70422 504-748-8937, 1-800-334-5109

J&K Schilling Nursery
26542 Mt. Pisgah Rd. Mt. Hermon, LA 70450 504-877-4251

Jackie's Garden Center
1318 Greenlaw Franklinton, LA 70438 504-839-9756

Lacombe Nursery & Landscaping
29062 Hwy 190 Lacombe, LA 70445 504-382-3668

Landscape Images Ltd.
655 Central Ave Jefferson, LA 70121 504-734-8380
Del; Bus.17yrs; 33%; Treeform Shrubs & specimen plants; Demo G; Exp; T Spade; Hours M – F 9–5

Larsen-Canone Nursery & Garden Center
21st Ave. Covington, LA 70433 504-893-4523

Laura's Green Thumb
59078 East Ave. Slidell, LA 70460 504-641-7363

Louisiana Landscape Garden Center
1701 Belle Chasse Hwy Gretna, LA 70056 504-393-6800

Louisiana Nursery Outlet
8680 Perkins Rd. Baton Rouge, LA 70810 504-766-0381

Louisiana Nursery Outlet
12289 Florida Blvd. Baton Rouge, LA 70815 504-766-0381

Louisiana Nursery Outlet
13121 Coursey Blvd. Baton Rouge, LA 70816 504-756-0008

Loupe's Nursery
310 Mar Dr. Lockport, LA 70374 504-532-5775

Martin's Garden Center Inc.
8527 Hwy 23 Belle Chase, LA 70037 504-393-4769
Del; Bus.12yrs; 33%; Demo G; Exp; Mem; Hours M-S 9-5:30, Sun 10-4

Marvin's Gardens
410 Red Cypress Rd. Patterson, LA 70392 504-395-5510

Mcmillan Nursery, Inc.
130 B Cumberland St. Bogalusa, LA 70427 504-735-7548

Nature's Creations
28105 Hwy 190 Ste. B Lacombe, LA 70445 504-882-2323

Nature's Greenhouse
1206 Magazine St. New Orleans, LA 70130 504-525-7000

Nine Mile Point Nursery Farms
2141 River Road New Orleans, LA 70121 504-436-4915

Nu-Earth Organics Inc.
3100 Veterans Memorial Blvd. Metairie, LA 70002 504-834-7888

Nys Nursery & Landscaping
10959 Honore Lane Baton Rouge, LA 70809 504-292-4171

Oak Haven Farms Garden Ctr.
18377 Blythe Rd. Prairieville, LA 70769 504-622-1058

Ott's Plants To Go
35526 Cane Market Rd. Denham Springs, LA 70706 504-665-7345

Pap's Garden & Gifts
11132 Hwy 21 South Bogalusa, LA 70427 504-730-5628

Paradise Gardens Of Baton Rouge, LLC
845 Lacrete Baton Rouge, LA 70810 504-756-0814

Paul Barrios Plants
226 Apple St. Norco, LA 70079 504-764-8258

Paula's Plants
3151 Emily Dr. Port Allen, LA 70767 504-749-3831

Plant World
6449 Hwy 90 East Morgan City, LA 70380 504-384-5211

Plants & Things Garden Center
34807 Hwy 16 / Watson   Denham Springs, LA 70706   504-667-3796

Plants 'N' Things
Ponchatoula, LA 70454   504-389-2029

Pretty Petals-N-Things
13700 L. Hwy 643   Vacherie, LA 70090   504-265-7000

Riley's Nursery & Greenhouse Inc.
Bogalusa, LA 70427   504-732-4958

Riverside Camellias
16068 Riverside Dr.   Covington, LA 70435   504-892-4259

Rose Garden Center
4005 Westbank Expwy   Marrero, LA 70072   504-341-5664
Bus 53yrs; 33%; Herbs, Perennials, Vegetables; Bonsai; Exp; Mem; Fert; Hours 9-6 M-S, 9-5 Sun.

Rosier's Nursery
Hwy 434   La Combe, LA 70434   504-892-1410

S&D PLants
14567 Frenchtown Rd.   Greenwell Springs, LA 70739   504-261-7255

Sand Dollar Nursery
61049 Military Rd.   Slidell, LA 70461   504-847-1153

Sandy Acres Nursery
150 Scott Dr.   Napoleonville, LA 70390   504-369-9878

Secret Gardens
2628 Metairie Rd.   Metairie, LA 70001   504-553-5526

Shademakers Nursery, Inc.
64754 I-55 Frontage Rd. N.   Arcola, LA   504-748-2847

Shadow Creek Azalea Farm
10218 Hwy 190 W   Hammond, LA 70401   504-345-8300

Slidel Discount Nursery
1605 Lawes St.   Slidell, LA 70461   504-641-8701

Southern Gateway
107 Timber Ridge St.   Belle Chasse, LA 70094   504-393-9300

Spencer Feed & Seed & Lawn & Garden
900 N. Columbia St.   Covington, LA 70433   504-892-4101

Sporl's Garden Center
801 E. St. BernaRd. Hwy   Chamlette, LA   504-271-3322

Nursery Specialties Key

Del - Delivery
Bus- Years in operation
% plant stock grown w/in 75 mile radius
Nat - Native plants
Exp - 2 years minimum staff experience
Mem - Member of nursery association
Learn C - Learning center
Fert - Organic fertilizer & pesticide
Demo G - Demonstration garden
News - Newsletter
T Spade - Tree spade

See page 68 for survey details.

Starke's Garden Center
4036 Hwy 311   Houma, LA 70360   504-872-6393

Sun Tropic Landscapes & Garden Center
1401 Whitney Ave.   Gretna, LA 70056   504-394-1166

Sunrise Nursery & Garden Center
248 Highpoint Blvd.   Denham Springs, LA 70726   504-664-2605

The Garden Trellis
8015 Maple St.   New Orleans, LA 70118   504-861-1953

The Palms Enterprise, Inc.
3944 Peters Rd.   Harvey, LA 70058   504-368-1788

The Plant Gallery
1135 Press   New Orleans, LA 70117   504-947-7554

Theriot & Brunet Nursery
639 Hwy 55   Montegut, LA 70377   504-594-4110

Tom Thumb Lawn & Garden Center
7346 Highland Rd.   Baton Rogue, LA 7808   504-767-5837

Uptown Garden Center LLC.
1521 S Jefferson Davis Pky   New Orleans, LA 70125   504-866-0075

Webre's Nursery
233 W. 2nd St.   Reserve, LA 70084   504-536-2359

Oklahoma

405 Area Code - Oklahoma City and Area

Badaway Farms
P.O. Box 508   Carney, OK 74832-9726   405-865-2784

Barnett Nursery, Landscape & Feed
17500 S. Sooner Rd.   Norman, OK   405-329-0061

Bent Tree Farm
Route 5 Box 636   Blanchard, OK 73010   405-387-5847

Breeze Hill Ranch
Route 3 Box 92e   Chandler, OK 74834-8519   405-258-4023

Caddo J Nursery
Rt. 1 Box 225y   Hinton, OK 73047   405-542-3554

Caesar's Greenhouse
2240 N. Air Depot Blvd.   Midwest City, OK 73141   405-424-5900

Chickasha Garden Center
1209 W. Grand Ave   Chickasha, OK 73018   405-222-4990

Coles Tree Farm
Rt. 1 Box 73AB   Lexington, OK 73051   405-872-3031

Contemporary Gardens
Rt. 5 Box 800   Blanchard, OK 73010   405-392-3485

Cooper Plant Mart Inc.
2534 N. McArthur   OKC, OK 73127   405-943-3322

Empress Farms
Hwy. 92 Yukon, OK 73099 405-350-0043

Ferguson Landscaping LLC, Coy
15301 S. Pennsylvania Ave. OKC, OK 73170 405-691-1980

Firstep
211 N. Walnut OKC, OK 73104 405-272-0366

Garden Center
2107 W. Main Norman, OK 73069 405-329-0800
Del; Bus.31yrs; Roses; Nat; Demo G; Exp; Mem; Learn C; Fert.

Gerdes Garden Center, Inc.
704 E. Independence Shawnee, OK 74804 405-273-7440

Goodman's Greenhouse
2801 E. Britton Rd. OKC, OK 73131 405-478-1198
Bus 39yrs; Annuals, perennials, and ground covers; Exp; Mem.

Grandpa's Greenthumb
6121 Nw 50th Bethany, OK 73122 405-789-9011

Hewitt's Retail Greenhouse
407 Leslie Drive Noble, OK 73068 405-872-5376

Holder's Herbs and Gifts
16287 S.E. 29th ave. Choctaw, OK 73020 405-390-2233
Bus. 17yrs; Herbs; Mem; Hours 10-5 Wed-Sat, 1-5 Sun, closed Mon-Tue.

Lacy's Flowers & Greenhouse
303 SE 2nd Street Lindsay, OK 73052 405-756-2696

Last Scent Farm
615 E. Robinson Norman, OK 73071 405-387-3795

Linda's Potting Shed
Hwy. 66E Chandler, OK 74834 405-258-5211

Marcum Greenhouses
169 N. Main Ave. Goldsby, OK 73093 405-288-2368

Marcum's Nursery Inc.
2121 Sw 119th St. OKC, OK 73093 405-691-9100

Martinbird Tree Farm
513 Ridgewood Dr. Tuttle, OK 73089 405-381-2910
Bus 20yrs; Nat; Balled and burlapped trees; Mem; Exp; T-spade;
Hours Tue-Sat 9-5

Mcloud Tree Farm
32301 Parkwood Road McLoud, OK 74851 405-706-4895

Meyer Tree Plantation Inc.
10010 Geronimo Dr. Norman, OK 73071 405-329-0019
Del; Bus 29yrs; Nat; Loblolly pines & root-control stock; Exp; Mem;
Hours Mon-Sat 8-4, Sun in season

Mid-America Nursery
7600 SW 119 OKC, OK 73173 405-691-6591

Moesel's HoRt. Haven
4315 NW 10th St. OKC, OK 73107 405-942-6996

Morrison Floral Company
4801 N. Meridian Ave. OKC, OK 73112 405-789-1622

Mother Earth
9211 Whispering Oaks Dr. Midwest City, OK 73130 405-733-3027

Nelson Nursery
7100 N. Frisco Road Yukon, OK 73099 405-354-5186

Nor-Mac Nursery
21250 E. Reno Harrah, OK 73045 405-654-2132

Northcutt's Greenlawn LLC
9501 Duffy Road Lexington, OK 73051 405-527-6675

Oasis Garden & Gift Shop
811 West 6th Ave Stillwater, OK 74074 405-624-7673
Del; Bus.6 yrs.; 33%, Perennials, herbs, unusual annuals; Exp;
DemoG; Bi-annual letter; Fert; Hours M-S 10-6 Sun 1-5

Paul B. Odom Jr.
7204 S. Pennsylvania Ave. OKC, OK 73159 405-685-5541

Pecan Specialities
7300 NW 16th OKC, OK 73127 405-789-6830

Pharaohs Gardens 13
121 E. 6 Street Glencoe, OK 74032 405-377-2553

Pioneer Nursery Inc.
7001 S. Council Road OKC, OK 73169 405-745-3036

Precure Garden Center
4535 NW 63Rd. St. OKC, OK 73132 405-789-4930

Precure Nursery
8125 W. Reno Ave. OKC, OK 73127 405-789-4930

Quality Plants Inc.
223 W. 32nd Ave. Stillwater, OK 74074 405-377-4494

Randy's Bonsai
2191 Harriet Blvd. Norman, OK 73072 405-364-4150
Bus 9yrs; 33%; Containerized Bonsai; Exp; Mem;
Hours Mon-Sat 12-5 & by appt; www.randysbonsaii.com

Redneck Herb & Garden
P.O. Box 208 Cement, OK 73017 405-489-3493

Sanctuary Water Gardens
Blanchard, OK
Del; Bus.7yrs; 33%; Nat; Aquatic & wetland plants; Demo G; Exp;
Mem; Hours W-Sat 9-6; Closed late Fall thru Winter except by appt.

TLC Nursery & Greenhouses
105 W. Memorial Rd OKC, OK 73114 405-751-0630
8208 N W Expressway OKC, OK 73126 405-720-0091
Del, 33%; Nat; Perennials Herbs Tropicals; Exp; Mem; Demo G;
Learn C; Fert; Hours M-S 8-6 Sun 12-5, extended hours Spring, Summer &
Xmas; www.tlcgarden.com

The Broker
P.O. Box 246 Anadarko, OK 73005 405-247-2077

Turning Leaf Garden Center
822 SW 134th Oklahoma City, OK 73170 405-912-7349

Warren and Son Nursery
405-769-2678
Del; Bus 50yrs; 33%; Nat; Perennials, Grasses Fruit Trees, Hostas; Exp;
Mem; News; Fert; Hours 8:30-5:30 M-Sat, closed Sun.

Whispering Springs Greenhouse
12647 SW 104th st. Mustang, OK 73064 405-376-2444
Bus 8yrs; 33%; Heat-loving plants; Exp; Mem; Spring hours- 8-6 M-F, 9-5 Sat, 12-3 Sun. Fall- 8-5 M-F, 9-12 Sat, 12-3 Sun.

580 Area Code - West, South and North Central Oklahoma

Abbott Nursery
Route 1 Box 390 Ada, OK 74820 580-332-2626

Ada Greenhouse
P.O. Box 1525 Ada, OK 74821 580-332-4500

Alexandria Nurseries
601 NE 15th Street Lawton, OK 73507 580-355-8871

Banks Greenhouse
Rt. 2 Box 72 Ada, OK 74820 580-332-5810

Bobsfarm Trees
P.O. Box 1069 Enid, OK 73709 580-234-6031

Bouldin's Sod Farm
P.O. Box 2145 Ponca City, OK 74602 580-762-6628

Brandt's Garden Center
P.O. Box 473 Boise City, OK 73933 580-544-3082

C&J Nursery
Rt. 1 Box 366 Elmore City, OK 73035 580-788-2376

Crenshaw's Flowers
909 Noble Street Alva, OK 73717 580-327-2149

Cimarron Garden Center
306 W. Willow Enid, OK 73701 580-237-3218

Dempsey Nursery
RR 1 Box 166B Blair, OK 73526 580-563-9194

Don's Plants
78 Candlewood Enid, OK 73701 580-233-0254

Dunn Nursery
2116 N. 5th Duncan, OK 73533 580-255-6614

Durant Wholesale Nursery Co.
Rt. 2 Box 187-5 Durant, OK 74701 580-924-4282

Nursery Specialties Key

Del - Delivery
Bus- Years in operation
% plant stock grown w/in 75 mile radius
Nat - Native plants
Exp - 2 years minimum staff experience
Mem - Member of nursery association
Learn C - Learning center
Fert - Organic fertilizer & pesticide
Demo G - Demonstration garden
News - Newsletter
T Spade - Tree spade

See page 68 for survey details.

Frederick Florist & Greenhouse
316 E. Brilliant Frederick, OK 73542 580-335-2712

Gardentown
1006 Scott Street Marlow, OK 73055 580-658-2111

Gleason's Instant Tree Co.
Rt. 2 Box 140K Rush Springs, OK 73082 580-476-3221

Grandpa's Greenhouse
Rt. 1 Box 210 Boswell, OK 74737 580-566-2463

Horn Canna Farm
Rt. 1 Box 94 Carnegie, OK 73015 580-637-2327

Idabel Florist & Greenhouse
614 S. Central Ave. Idabel, OK 74745 580-286-3373

J&J Lakeview Farm
Route 1 Box 92AB Roosevelt, OK 73564 580-639-2791

Keathly's Nursery/Landscape Co.
2448 Bridge Ave. Ponca City, OK 74604 580-762-2922

Larry's Poor Boy Plants
Box 214-Cleo Springs, OK 73729 580-438-2228

Leon's Greenhouse
P.O. Box 578 Kingston, OK 73439 580-564-2166

Ligon's Garden Center Inc.
P.O. Box 245 Duncan, OK 73534 580-255-6087

Mary's Green House
621 E. Florida Ave. Waurika, OK 73573 580-228-2124

McKendrick Greenhouse & Farm
P.O. Box 5 Caney, OK 74533 580-889-9173

Me & Mom's Watergarden
216 Shawna Drive Durant, OK 74701 580-920-0102

Mountain View Iris Gardens
Box 228E Apache, OK 73006 580-492-5183

Old Roses & Such
1009 N. Illinois Weatherford, OK 73096 580-772-1095

Olde Town Nursery
Rt. 3, Box 152B Altus, OK 73521 580-482-6007
Del; Bus 5yrs; Nat; Exp; Mem; Demo G; Mearn C; Fert; Hours Winter M-F 9-5, Sat 9-2, closed Sun. Other seasons 9-6 M-S, 12-5 Sun.

Packers Plants Inc.
1407 College Blvd. Alva, OK 73717 580-327-3522

Pear Tree Nursery
P.O. Box C Carnegie, OK 73015 580-654-2488

Plants & Plenty
622 E. Camelback Road Duncan, OK 73533 580-255-1213

Push 'N' Herbs
P.O. Box 87 Moyers, OK 74557 580-298-6559

Red Hill Nursery & Gifts
RR 2 Box 42 Cheyenne, OK 73628 580-497-2118

Red Oak Nursery
907 Patterson | Comanche, OK 73529 | 580-439-6000

Rhodes Greenhouse
Rt. 1 Box 127-1 | Durant, OK 74701 | 580-920-0703

Riley's Iris & Daylily Garden
Rt. 1 Box 125 | Wakita, OK 73771 | 580-594-2450

Sunshine Nursery
North of Clinton | Clinton, OK | 580-323-6259
Bus 20yrs; 33%; Nat; Variety of trees and natives; Exp; Mem; Demo G; Fert; Hours 8-5 M-F, 8-12 Sat, closed Sun. Spring hours extend 8-5, closed Sun. www.sunshinenursery.com

The Plant Place
501 W. 2nd St. | Watonga, OK 73772 | 580-623-8579

918 Area Code - Tulsa & Northeast Oklahoma

A New Leaf
2306 S. One Place | Broken Arrow, OK | 918-451-1491

Adate Country Orchard Co., Inc.
Route 2 Box 2075 | Stilwell, OK 74960 | 918-695-7333

Allton Plant Farms
P.O. Box 366 | Jenks, OK 74037 | 918-299-5434

Amanda's Greenhouse
9503 W. 162nd St. South | Sapulpa, OK 74066 | 918-321-3757

Ann's Daylilies & Perennials
13218 N. 95th E. Avenue | Collinsville, OK 74021 | 918-371-6035

Art By Nature
1127 S. Louisville | Tulsa, OK 74112-5119 | 918-835-4853

Azelea Gardens Nursery
6310 E. Archer | Tulsa, OK | 918-835-5994

B & J's Greenhouses
Route 4 Box 112 | Muldrow, OK 74948 | 918-427-5697

Baase Garden Center Inc.
P.O. Box 122 | Claremore, OK 74018 | 918-341-3455

Bauman's Greenhouse
25695 E. Hwy. 51 | Broken Arrow, OK 74014 | 918-486-6298

Bogarts Garden Center & Nursery
406 Hickory Dr. | Henryetta, OK 74437 | 918-652-4000

Bookout's Nursery
P.O. Box 102 | Henryetta, OK 74437 | 918-652-2238

Brewster's Produce
P.O. Box 155 | Whitefield, OK 74472 | 918-967-8562

Brown Farms LLC
15923 E. 116th Street N. | Collinsville, OK 74021 | 918-274-9800

Buckaroo Nursery
5229 W. Okmulgee St. | Muskogee, OK 74401 | 918-682-1577

Bucky's Naturals
Route 5 Box 1184 | Muldrow, OK 74948 | 918-427-6460

BV Nursery & Gifts
Route 1 Box 213B | Cleveland, OK 74020 | 918-243-5202

Calvert Tree Farm Inc.
51st & Coweta Road S | Broken Arrow, OK 74011 | 918-455-5041

Chigger Hill Canna Farm
Rt. 2 Box 2080 | Porum, OK 74455 | 918-484-2490

Cox & Rossi
4951 N. Lewis | Tulsa, OK 74130 | 918-425-1117

Cymarron Valley Greenhouses #1
Rt. 3 Box 2360 | Cushing, OK 74023 | 918-225-5172

Eastern Oklahoma State College
1301 W. Main St. | Wilburton, OK 74578 | 918-465-5770

Edwards Floral & Garden Center
303 S. Mickey Mantle Bl. | Commerce, OK 74339 | 918-675-4334

Eucha Gardens & Specialty
P.O. Box 1414 | Jay, OK 74346 | 918-253-6008

Evans Greenhouse & Nursery
111 Beck Street | Bartlesville, OK 74006 | 918-333-3248

Ghost Hollow Farms
Rt. 1 Box 2140 | Ripley, OK 74062-9736 | 918-372-4612

Grandview Nursery
901 Woodard Road | Tahlequah, OK 74465 | 918-456-4010

Green Acre Sod Farms
12300 S. Mingo Road | Bixby, OK 74008 | 918-396-5918

Green Country Gardens Inc.
RR 1 Box 245 | Tahlequah, OK 74464 | 918-655-6022

Green Thumb Nursery
4605 Nowata Road | Bartlesville, OK 74006 | 918-333-4091

Green Valley Nursery
17102 E. 430 Road | Claremore, OK 74017 | 918-341-3030

Holiday Farms Of Oklahoma LLC
P.O. Box 326 | Adair, OK 74330 | 918-785-4138

Honey Valley Nursery Inc.
1020 E. Lincoln | Sapulpa, OK 74066 | 918-224-1848

Huntree Nursery And Christmas Tree Farm
Hwy 59 & Hwy 141 | Pocola , OK 74902 | 918-775-7800

Juniper Hill Farm, Inc.
9740 E. 121st St. | Bixby, OK 74008 | 918-369-5763
Del; Bus 30yrs; Nat; Landscape plants & unusual; Demo G; Exp; Learn C;

Kendall Plant Sales
2629 E. 8th St. | Tulsa, OK 74104 | 918-585-1489

Kunze Farms
Route 1 | Porter, OK 74454 | 918-357-2007

Manns Greenhouse & Flower Shop
224 S. 3rd st. | McAlester, OK 74501 | 918-423-6403
Del; Bus.72yrs; 33%; Nat; Exp; Mem; Fert.

Mingo Turf & Tree Co.
161st St. S. & Mingo Rd. Bixby, OK 74008 918-366-8873

Muskogee Farmers Market
2783 W. 83Rd. St. S. Muskogee, OK 74401 918-687-4993

Nickel Creek Nursery
8718 S. 33Rd. W. Avenue Tulsa, OK 74132 918-446-4088
Del; Bus.3yrs; 33%; Clematis vines; Exp; Fert; Hours M-Sat 9-6

Oklahoma Home Grown Plants
Route 1 Box 372 Kiowa, OK 74553 918-432-6112

Potter Landscaping & Nursery
222 S. Elm Street Miami, OK 74354 918-540-2124

Pryor Creek Nursery
HC 65 Box 20-4 Pryor Creek, OK 74361 918-825-6121

Raska Nusery, Inc.
211 E. Mason Drive Broken Arrow, OK 74012 918-251-3329

River Bend Nursery Inc.
10833 S. Delaware Ave. Tulsa, OK 74137 918-299-6081

Riverview Sod Ranch
P.O. Box 100 Leonard, OK 74043 918-366-4141

Rockwood Gardens
3305 Racetrack Rd. Pocola, OK 74902 918-436-0100
Del; Bus.4 yrs; Nat; Exp 2 yrs, Fert; TSpade; Hours 9-5; Landscape

Stringer Nursery & Garden
100020 E. 91st street Tulsa, OK 74133 918-250-0554
7203 E. 41st street Tulsa, OK 74145 918-628-0296
Bus.44yrs; Nat; Azaleas, Roses, Landscape plant material; Mem; Fert.

The Garden Shop
201 S. Grant Wagoner, OK 74467 918-485-3842

The Herb Garden Plus
Route 4 Box 180 Whitefield, OK 74472 918-967-2782

Toedt's Garden/Landscape for Better Living
1321 W. 41st Street Tulsa, OK 918-445-3535

Tree House Nursery And Plant Farm
Hwy 271 W. Of Pocola, OK 74902 918-436-2377

Wendland Nursery & Landscaping
9301 Garnett Rd. Tulsa, OK 918-252-4214
Del; Bus.32yrs; 33%; Nat; Large Trees, B&B stock; Demo G; Exp; Mem; Learn C; Fert; Tree spade; Hours m-sat 8-5, Sun 12-5 seasonal

Nursery Specialties Key

Del - Delivery
Bus- Years in operation
% plant stock grown w/in 75 mile radius
Nat - Native plants
Exp - 2 years minimum staff experience
Mem - Member of nursery association
Learn C - Learning center
Fert - Organic fertilizer & pesticide
Demo G - Demonstration garden
News - Newsletter
T Spade - Tree spade

See page 68 for survey details.

Texas

806 Area Code - Amarillo, Lubbock and Areas

Ag Products Incorporated
P.O. Box 786 Levelland, TX 79336 806-894-6665

Ansel Family Farm Stores Inc
501 N Pumas Ave Dumas, TX 79029 806-935-7719

Bert Thomas Lawn And Landscape Inc
11701 Truman St Amarillo, TX 79118 806-622-3135

Cade Greenhouse
Route 1 Box 66 Slaton, TX 79364 806-828-6464

Canyon Garden Center
417 17th St Canyon, TX 79015 806-655-4880

D And D Gardens
401 Miller St Spur, TX 79370 806-271-3237

Double E Cox Tree Farm
RR 2 Box 13 Amarillo, TX 79088 806-995-4417

Family Tree Nursery And Tree Farm
99th And Quaker Lubbock, TX 79416 806-794-2601

Garden Center
P.O. Box 521 Dumas, TX 79029 806-935-2894

Garden Patch
711 Hwy 87 South Dalhart, TX 79022 806-249-5212
Del; Bus.3yrs; Nat; Exp; Demo G; Learn C; Fert; Spring M-Sat 9-9 Sun 1-6

Garden Pleasures
10502 University Lubbock, TX 79423 806-748-9597

Gardener's Outlet
1805 N Frankford Ave. Lubbock, TX 79416 806-744-5404

Gerry's Greenhouse
905 N Lynn Ave Lamesa, TX 79331 806-872-8660

High Plains Natural Garden
1230 S Bonham St Amarillo, TX 79102 806-376-6948

Holland Gardens
4315 50th St Lubbock, TX 79414 806-792-6336

Ivy Gardens
1318 Municipal Dr. Lubbock, TX 79403 806-744-4839

Knight's Nursery
P.O. Box 528 Canadian, TX 79014 806-323-5577

Kristy's Plants
710 East American Blvd Muleshoe, TX 79347 806-272-5536

Little Red Riding Hood Nursery
4006 34th Street Lubbock, TX 79410 806-795-4834
3839 34th Street Lubbock, TX 79410 806-799-8938

Love & Son Nursery
1103 Ross St. Amarillo, TX 76012 806-373-9563

Mary Lee Gardens
6702 82nd St Lubbock, TX 79424 806-798-1444
Del; Bus. 1yrs; 33%; Nat; Exp; Mem; Fert; Hours 9-6 M-Sat. closed Sun.

McLain Lawn And Garden Center Inc
1302 West Park Avenue Hereford, TX 79045 806-364-3300

Panhandle Greenhouse And Floral Inc
11900 S FM 1541 Amarillo, TX 79118 806-622-2683

Pollies Country Garden
Rt. 2 Perryton, TX 79070 806-435-3909

Roberts Tree House
2503 Dimmitt Road Plainview, TX 79072 806-293-2613

Shafer Nursery
P.O. Box 514 Muleshoe, TX 79347 806-272-4653

Sparkman Nursery
11109 Slide Rd Lubbock, TX 79424 806-794-3614
Native Oak Trees

Taylor Rose Garden And Gifts
3400 S Hope Rd Amarillo, TX 79124 806-352-0007

Texas Garden's
P.O. Box 597 Kress, TX 79052 806-684-2375

TG Trees
Route 4 Box 175 Lubbock, TX 79424 806-794-6766

The Flower Bed C/O Valerie Sturgess
HCR 2 Box 119 Happy, TX 79042 806-627-4486

The Garden Patch
1311 Alcove Ave Lubbock, TX 79416 806-793-0982

The Greenhouse
P.O. Box 803 Borger, TX 79008 806-274-5666

The Greenhouse - Carolyn Hartley
135 Marshall Howard Littlefield, TX 79339 806-385-3406

Tom's Tree Place
5104 34th St. Lubbock, TX 79404 806-799-3677

512 Area Code - Austin and Area

Arbormaster Tree Farm
3702 RR 620 South Austin, TX 78734 512 -263-9214

Aus Tex Garden Supply
5101 Burnet Road Austin, TX 78756 512 -459-7311

Austin Landscape And Bldg Supplies
16407 FM 1325 Austin, TX 78728 512 -255-2517

Austin Nursery And Landscape- David Yepez
P.O. Box 1116 Pflugerville, TX 78660 512 -251-0190

Barton Springs Nursery Co
3601 Bee Caves Rd Austin, TX 78746 512 -328-6655

Bastrop Gardens
316 Old 71 Cr 335 Cedar Creek, TX 78612 512 -303-5672

Big Bear Nursery
1509 Falcon Oaks Leander, TX 78641 512 -260-0350

Bloomers Garden Center
1074 Hwy 290 Elgin, TX 786210000 512 -285-9445

Bolton Works Nursery
P.O. Box 1100 Dripping Springs, TX 78620 512 -894-4234

Bonsai Exotica
16400 S. IH-35 Buda, TX 78610 512-312-1614
Bonsai trees & starter plants; Learn C; Demo G; Hours W-Sun 1-5:30, call for other hours.

Bonsai Garden Of TX
RR1, Box 564-X Cedar Creek, TX 78612 512 -321-3379

Breed And Company Inc
718 West 29th St Austin, TX 78705 512 -474-6679

Bunny Bear Hill Farms
15255 FM 150 W Driftwood, TX 78619 512 -858-7653

Carlois Smith Landscaping And Nursery
5351 Sherri Ann Rd San Antonio, TX 78233 512 -637-1532

Cedar Park Turfgrass And Garden Inc
1715 FM 1431 Cedar Park, TX 78613 512 -250-1885

Centex Trees And Shrubs
Rt. 5 Box 70 Bastrop, TX 78602 512 -303-3765

Courtyard Gardens- Bud Twilley
2729 Exposition Blvd #114 Austin, TX 78703 512 -708-1640

Evergreen Farms Christmas Trees
242 Monkey Rd Elgin, TX 78621 512- 281-4833

Family Gardens- George J Strmiska Jr
101 Cr 384 Bartlett, TX 76511 512- 859-2378

Flowerbeds And Watergardens
P.O. Box 35 Liberty Hill, TX 78642 512- 515-6700

Forest Grove Nursery #2
Rt. 1 Box 247 Rockdale, TX 76567 512- 446-2995

Garcias Nursery
442 Commercial St Eagle Pass, TX 78852 512- 773-3367

Garden Ville Of San Marcos
2212 RR 12 San Marcos, TX 78666 512- 754-0060

Gardens Of The Ancients
4800 Yager Ln Manor, TX 78653 512- 272-9062
33%; Fert; Medicinal herbs; call for hours

Gardens Unique
13420 Cedar Lime Rd Leander, TX 78641 512- 267-0455

Gardner's Nursery
1209 W Anderson Ln Austin, TX 78757 512- 452-2610

Garza's Greenhouse
421 South Guadalupe San Marcos, TX 78666 512- 392-2756

Geo Growers
6300 C Hwy 290 West Austin, TX 78735 512-892-2722

Great Hills Garden Center
6914 McNeil Dr. Austin, TX 78729 512-219-9600

Greenhouse Nursery And Landscaping, The
5478 F M 2673 Canyon Lake, TX 78133 512- 899-4642

Harmony Gardens
10814 N Lamar Blvd Austin, TX 78753 512 -835-4740

Heirloom Gardens
2700 W Anderson Ln., Ste 904 Austin, TX 78757 512 -451-0241

Henna Nursery
735 Louis Henna Blvd. Austin, TX 78664 512-733-1952

Heritage Design Landscaping
5205 Hudson Bend Road, Austin, TX 78734 512-266-7873
Del; Bus.5 yrs; 33%; Nat; Southwest Natives, Desert Plants, Deer Resistant Plants; Demo-G; Fert; Hours M-F 9-5, Sat 9-4, closed Sun.
www.6200nune.com or www.jillsnursery.com

Hernandez Nursery And Landscaping Service
P.O. Box 66 Donna, TX 78537 512- 464-7451

Hill Country Garden Center
13561 Pond Springs Rd. Austin, TX 78720 512-258-0093

Hill Country Home And Garden
4319 Hazy Hills Dr Spicewood, TX 78669 512 -264-3432

Hill Country Water Gardens And Nursery
1407 N Bell Blvd Hwy 183 Cedar Park, TX 78613 512 -260-5050

Horticulture Center
11606 Sunset Drive Austin, TX 78748 512 -282-1011

Howard Nursery
111 E. Koenig Ln. Austin, TX 78751 512-453-3150

It's A Jungle
907 Kramer Ln. Austin, TX 78758 512-453-3150

It's About Thyme
11726 Manchaca Rd Austin, TX 78748 512 -280-1192
Del; Bus. 22 yrs; 33%; Nat; Natives Antique Roses, herbs and perennials; Mem; Demo G; Learn C; Fert; Open every day Spring hours 8-6, Winter 9-5 ; www.itsaboutthyme.com

Jade Gardens
12404 Ranch Road 12 Wimberley, TX 78676 512 -847-2514

Koi Creek Nursery
15401 RR 12 Ste. 101 Wimberly, TX 78676 512- 847-1239

Nursery Specialties Key

Del - Delivery
Bus- Years in operation
% plant stock grown w/in 75 mile radius
Nat - Native plants
Exp - 2 years minimum staff experience
Mem - Member of nursery association
Learn C - Learning center
Fert - Organic fertilizer & pesticide
Demo G - Demonstration garden
News - Newsletter
T Spade - Tree spade
See page 68 for survey details.

Lady Bird Johnson Wild Flower Ctr.
4801 LaCrosse Avenue Austin, TX 78739 512 -292-4200

Living Desert Cactus Nursery
12719 W Hwy 71 Austin, TX 78736 512-263-2428

Lost Pines Nursery
Route 3 Box 410a Bastrop, TX 78602 512- 321-4175

Maldonado Nursery And Landscape
120 E Zipp Road New Braunfels, TX 78130 512 -625-1020

Masters Greenery
906 Highway 90 West Castroville, TX 78009 512- 538-3001

McIntire's Garden Center Inc
303 Leander Rd. Georgetown, TX 78626 512 -863-8243

Monodale Farms Vineyard And Nursery
200 CR 133 Hutto, TX 78634 512 -759-5126

Murffy's Nursery
901 Sam Bass Rd. Round Rock, TX 78681 512-255-3353

Natural Gardener
8648 Old Bees Cave Rd Austin, TX 78735 512-288-6113

Neal's Nursery
126 W. Slaughter Austin, TX 78709 512 -282-8199

Ned's Nursery
P.O. Box 532227 Harlingen, TX 78553 512 -423-2154

Newton Nursery Austin
849 W 26th St. Houston, TX 77008 512 -310-0055

Olde Thyme Gardens
950 CR 365 Taylor, TX 76574 512 -352-7673

Park Place Gardens Inc
2710 Hancock Dr. Austin, TX 78731 512 -458-5909

Piddler Creek Nursery
1201 Central, Tx Expwy. Lampasas, TX 76550 512 -556-6800

Pots And Plants
5902 Bee Cave Rd. Austin, TX 78746 512 -327-4564

Red Barn Garden Center
13858 Hwy 183 N. Austin, TX 78750 512 -335-0122

Rhapis Gardens
P.O. Box 287 Gregory, TX 78359 512-643-2061
Bus 26yrs; Many palms, grape ivy & ming aralias;
www.Rhapisgardens.com

Rosemont Nursery
707 W 4th St Lampasas, TX 76550 512 -556-2021

Sledd Landscape Nursery
1211 West Lynn Austin, TX 78703 512 -478-9977

Snoopers Nursery
11304 N IH-35 Austin, TX 78753 512 -836-3458

Southside Nursery
P.O. Box 2063 Georgetown, TX 78626 512- 863-7798

Sunshine Landscape & Garden Center
2000 RR 620 South Austin, TX 78734 512-263-5275

Sunshine Nursery
13105 Canyon View Leander, TX 78641
Call for hours. Cactus & succlents

Sylvania Nursery
455 Allison Rd. Austin, AR 72007 512-843-7323

Tanglewood Plants
Route 2 Box 694A Lexington, TX 78947 512- 446-2268

Tangram Nursery Co
5315 Tangram Nursery Rd Maxwell, TX 78656 512- 396-0667

Ted's Trees
1118 Tillery St. Austin, TX 78702 512-836-9001

Texas Landscape Nursery
P.O. Box 33910 San Antonio, TX 78265 512 -646-6666

Texzen Gardens
4806-Burnet Rd. Austin, TX 78756 512- 454-6471

The Emerald Garden
6910 Highway 71 West Austin, TX 78735 512 -288-5900

The Flower Patch Nursery
P.O. Box 866 Georgetown, TX 78628 512 -869-2678

The Garden House Nursery
3310 N Main St. Taylor, TX 76574 512 -352-2494

The Great Outdoors Landscape & Nursery
2730 Congress Av Austin, TX 78704 512-448-2992
Del; Bus.15yrs; 33%; Nat; Palms, tropical-looking; Exp; Demo G; Learn C; News; Fert; www.gardenadventures.com

Tron Gardens- Ronald C Miller
9802 Vikki Terrace Austin, TX 78736 512 -288-2442

Valley Garden Center Inc
P.O. Box 5429 McAllen, TX 78502 512- 682-9411

Velas Nursery
108 North Nebraska San Juan, TX 78589 512- 787-7617

Walton's Florist & Nursery
5604 Bee's Cave Rd. Austin, TX 78746 512-327-1206

Wolfe Nursery
1905 S. Capital of TX Hwy Austin, TX 78746 512-306-8877
8701 Research Blvd. Austin, TX 78758 512-339-6268
4715 S. Lamar Blvd. Austin, TX 78745 512-892-4926
1900 S. Bell Blvd. Cedar Park, TX 78613 512-331-9510
901 Sam Bass Rd. Round Rock, TX 78681 512-255-3353
www.wolfenursery.com

Wyatt's Nursery
6302 W. Hwy 290 Austin, TX 78735 512-892-5197

830 Area Code - Hill Country Area

Angel's Nursery
P.O. Box 333 Lytle, TX 78052 830-709-5602

Barkleys Home And Ranch Center
32450 IH 10 West Boerne, TX 78006 830-249-2656

Blanco Gardens
P.O. Box 1556 Blanco, TX 78606 830-336-2671

Boerne In Bloom Garden Center Inc
123 E Frederick St Boerne, TX 78006 830-249-7121

Collins Nursery
P.O. Box 1000 Medina, TX 78055 830-589-2877

Country Gardens And Seed
115 East Cardwell Uvalde, TX 78801 830-278-2204

Dodds Family Tree Nursery Incorporated
515 West Main St Fredericksburg, TX 78624 830-997-9571

Dunns Green Thumb
6932 US Hwy 181 N Floresville, TX 78114 830-253-2401

Gordon's Gardens Nursery
P.O. Box 210 Utopia, TX 78884 830-966-3440

Green Acres Nursery Incorporated
Rt. 1 Box 178, Hwy. 97 Gonzales, TX 78629 830-672-3277

Greenland Nursery
411 North Teel Devine, TX 78016 830-665-2193

Henry's Nursery
1795 FM 3443 Eagle Pass, TX 78852 830-773-0615

Hill Country African Violets
32005 IH 10 West Boerne, TX 78006 830-249-2614

Hill Country Gardens
414 S Orange Fredericksburg, TX 78624 830-997-3064

Laguna Nursery Inc
314 West Main Street Uvalde, TX 78801 830-278-6769

Landscape Solutions And Nursery
3045 US Hwy 90 E Castroville, TX 78009 830-762-3959

Lani's Garden Delights
28024 Indian Path New Braunfels, TX 78132 830-438-2860

M G Nursery And Landscaping
21151 FM 471 S Natalia, TX 78059 830-663-5536

Maldonado Nursery And Landscape
3603 Memorial Blvd. Kerrville, TX 78028 830-896-6869

Maldonado Scenic Landscaping & Nursery
136 Adler Rd. Boerne, TX 78006 830-249-4694

Maxine's Plants And Landscaping
800 Forest Tr. Bandera, TX 78003 830-460-8283

Medina Garden Nursery
Rt. 16 Box 107 Medina, TX 78055 830-589-2771

On The Farm Inc
4452 FM 536 Pleasanton, TX 78064 830-393-4944

Orchids And Roses
604 N. 1st Carrizo Springs, TX 78834 830-876-2447

Plantiques
26791 Hwy 281-N | San Antonio, TX 78260 | 830-980-8224

Rocking R Nursery Inc
445 Private Rd. 8453 | Kenedy, TX 78119 | 830-583-3222

Schultz Nursery
100 W. Hubinger | Marion, TX 78124 | 830-914-2384

Schulz Nursery
P.O. Box 128 | Marion, TX 78124 | 830-914-2384

Schumacher's Hill Country Gardens, Inc.
588 Highway 1863 | New Braunfels, TX 78132 | 830-620-5149

Serrengetti Gardens
2217 State Hwy 173 N. | Devine, TX 78016 | 830-663-5501

Shirley Lynn Nursery
548 Hwy 239 | Kenedy, TX 78119 | 830-583-3151

Sunshine Garden Center
P.O. Box 1045 | Bracketville, TX 78832 | 830-563-2610

Tanneberger's Nursery
2102 Memorial Blvd. | Kerrville, TX 78028 | 830-896-4094

The Greenery
1245 Bandera Hwy | Kerrville, TX 78028 | 830-896-7553
Del, Bus.18yrs; 33%; Nat; Natives; Demo G; Exp; Mem; Learn C; Fert; News;Hours M-F 7:30-5:30pm, Sat 8-2pm; E-mail: greenery@ktc.com

The Little Acre Nursery
3708B Ave. F | Del Rio, TX 78840 | 830-774-7030

The Plant Haus Nurseries Inc
956 North Walnut | New Braunfels, TX 78130 | 830-629-2401

The Potting Shed
P.O. Box 519 | Waelder, TX 78959 | 830-778-7272

Villarreals Instant Trees
931 CR 679 | Natalia, TX 78059 | 830-772-3312

409 Area Code - Beaumont, Galveston, College Station and Area

All In One Nurseries And Landscaping
2701 Broadway | Galveston, TX 77550 | 409-765-6447

Arbor Ridge Tree Farm
FM 3090 | Navasota, TX 77868 | 409-825-7778

Nursery Specialties Key

Del - Delivery
Bus- Years in operation
% plant stock grown w/in 75 mile radius
Nat - Native plants
Exp - 2 years minimum staff experience
Mem - Member of nursery association
Learn C - Learning center
Fert - Organic fertilizer & pesticide
Demo G - Demonstration garden
News - Newsletter
T Spade - Tree spade
See page 68 for survey details.

Arena's Nursery And Landscape
1165 Monterrey | Beaumont, TX 77706 | 409-842-1687

ARLDT Nursery
P.O. Box 471 | Giddings, TX 78942 | 409-542-2902

B And B Nursery
9675 Hwy 190 West | Livingston, TX 77351 | 409-646-4944

B B's Garden
P.O. Box 11 | Chappell Hill, TX 77426 | 409-836-0250

Beaumont Greenery
3925 Delaware St | Beaumont, TX 77706 | 409-898-7740

Beech Creek Farm
P.O. Box 39 | Wren, TX 77664 | 409-547-2235

Bloomer's Garden Center
1245 Loop 304 East | Crockett, TX 75835 | 409-546-0717

Boyett's Plant Farm
Rt. 3 Box 2700 | Bronson, TX 75930 | 409-586-3191

Brazosport Green Thumb Nursery Inc
1071 N Brazosport Blvd. | Richwood, TX 77531 | 409-266-7760

Briarfield Gardens
219 FM 954 | Round Top, TX 78954 | 409-249-3973

Bridge Valley Nursery
427 West Travis | La Grange, TX 78945 | 409-968-6093

Bucket Brigard Herb And Garden Center
P.O. Box 609 | Sour Lake, TX 77659 | 409-287-3754

Burkett Nursery
1828 Willow | Liberty, TX 77575 | 409-336-8177

Bushes & Blossoms Landscaping Inc
2930 Loop 197 | TX City, TX 77590 | 409-945-9840

Bushes, Blossoms & Bouquets 2
12531 FM 1764 | Santa Fe, TX 77510 | 409-925-2090

Bushy Tree Nursery
P.O. Box 582 | Sealy, TX 77474 | 409-885-6719

Carters Florist Nursery And Landscaping
1416 South Frazier Street | Conroe, TX 77301 | 409-756-7222

Central Heights Nursery And Pro
Route 13 Box 2820 | Nacogdoches, TX 75961 | 409-564-1396

Changing Seasons Lawn And Garden Ctr.
7067 Pleasure Lake Dr. | Willis, TX 77318 | 409-890-2153

Circle W Ponds
Rt. 3 Box 216 | Kirbyville, TX 75956 | 409-423-6757

CJ Grass Nursery
P.O. Box 252 | Groveton, TX 75845 | 409-642-1660

Cliff Cook Jr. Nursery
8165 Gladys | Beaumont, TX 77706 | 409-861-3400

Community Orchard And Garden Center
Rt. 1 Box 283 A | Wharton, TX 77488 | 409-532-0185

Conroe Garden Center
3301 W Davis | Conroe, TX 77304 | 409-756-5569

Contemporary Landscape Service & Nursery Inc.
106 North Ave. | Bryan, TX 77801 | 409-846-2775

Cooks Nursery
Route 14 Box 4150 | Nacogdoches, TX 75964 | 409-564-6359

Cook's Nursery
1424 Nederland Ave | Nederland, TX 77627 | 409-724-2665

Country Gardens
108 Kessler | Schulenburg, TX 78956 | 409-443-4648

Cross Roads Garden Center
1602 S Market | Hearne, TX 77859 | 409-279-3061

Crosstown Nursery
412 Hwy 365 | Port Arthur, TX 77640 | 409-736-9191

Crystal Beach Plant Farm
P.O. Box 2109 | Crystal Beach, TX 77650 | 409-684-5723

Cuernavaca Nursery
509 S First St | Conroe, TX 77301 | 409-539-2671

Cypress Pointe Nursery And Landscaping
325 Old Hwy 90 W | Vidor, TX 77662 | 409-783-9898

D And R Garden Center Inc.
1248 9th Ave | Port Arthur, TX 77640 | 409-985-4416

Dee's Nursery And Lawn Ornaments
Rt. 1 Box 316 D | Burkeville, TX 75932 | 409-565-2529

Diane's Produce And Nursery
2380 Hwy 19 N | Huntsville, TX 77340 | 409-295-6747

Discount Trees Of Brenham
2800 N. Park St | Brenham, TX 77833 | 409-836-7225

Dixie Nursery And Gift
1320 Hwy 1131 | Vidor, TX 77662 | 409-786-1368

D'Lana Farms
P.O. Box 8532 | Lumberton, TX 77657 | 409-755-1452

Earma Lee Collins Nursery
1316 Cottonwood | Huntsville, TX 77340 | 409-295-7707

Ecotone Gardens
P.O. Box 1445 | Kountze, TX 77625 | 409-246-3070

Edgebrook Farms Nursery Inc
322 E Main St | Bellville, TX 77418 | 409-865-9925

El Kay Farms And Nursery Inc
18451 Kinkaid Rd. E | Montgomery, TX 77316 | 409-597-6107

Emerald Woods Greenhouses
36 Lakeview Dr | New Waverly, TX 77358 | 409-344-2240

Evergreen Nursery
600 E Henderson | Angleton, TX 77515 | 409-849-6330

Everlings Nursery
5310 Avenue Q | Galveston, TX 77550 | 409-744-1767

Exterior Accents
21 River Run | Brazoria, TX 77422 | 409-798-5788

Fischer's Nursery
Route 1, Box 59 | Giddings, TX 78942 | 409-366-2416

Florence's Nursery
Route 1, Box 203C | Buna, TX 77612 | 409-994-2084

Forever Green Nursery
2510 N Mechanic | Elcampo, TX 77437 | 409-543-1318

Garcia's Nursery
301 Emmett St | Beaumont, TX 77701 | 409-832-0632

Garden Center
3195 Concord | Beaumont, TX 77703 | 409-835-5933

Garden Of Eden Nursery
3333 Highway 332 | LakeJackson, TX 77566 | 409-798-8416

Green Glen Nursery
5849 25th Street | Groves, TX 77619 | 409-962-2135

Greenhouse Nursery
1327 North Alabama Rd | Wharton, TX 77488 | 409-532-0774

Gulf Coast Nurseries
302 32nd St | Galveston, TX 77550 | 409-762-8733
Del; 33%; Nat; Tropicals; Exp; Mem, Demo G, Lern C; News,
Fert; Hours M-Sat 9-5pm, Sun 10-4; www. gulfcoastnursery.com

Gulf Coast Nursery And Landscaping
302 32nd St. | Galveston, TX 77550 | 409-765-8733

Hernandez Nursery
1004 West Loop | El Campo, TX 77437 | 409-543-6327

Hilltop Nursery
2508 Highway 69 North | Lufkin, TX 75901 | 409-637-7205

Hummingbird Farms
8273 South Killarney Dr | Beaumont, TX 77705 | 409-722-6454

Hunt Reeves Nursery
Rt. 3 Box 85 | Buna, TX 77612 | 409-994-2921

Ivy Nursery Incorporated
2202 West Houston | Jasper, TX 75951 | 409-384-9077

J & L Bluebonnet Plantation
Highway 6 | Navasota, TX 77868 | 409-825-4140

J and S Plants
P.O. Box 391 | Midway, TX 75852 | 409-348-6836

JD's Pots & Plants
Route 1, Box 59 | Apple Springs, TX 75926 | 409-831-2986

Jean's Garden And Greenhouse
781 N Harris | Giddings, TX 78942 | 409-542-6225

Joe's Green Thumb Nursery
Rt. 4, Box 50-J | Dayton, TX 77535 | 409-258-7202

Jones' Nursery
P.O. Box 604 Deweyville, TX 77614 409-746-2332

Joshua Tree Nursery
5565 Roundbunch Orange, TX 77630 409-738-2070
Del; Bus.4yrs; Nat; Exp; Demo G; Fert.

Junior's Garden
2315 45th St. Galveston, TX 77550 409-762-4557

Kenefick Nursery And Garden Center
Route 5 Box 535 Dayton, TX 77535 409-258-8637

Kim's Flower And Nursery
6613 W Port Arthur Rd. Port Arthur, TX 77640 409-736-2465

Kim's Plants
P.O. Box 351 Gause, TX 77857 409-280-5197

La Selva Orchids And Nursery
3910 Harvey Road College Station, TX 77845 409-774-4776

Lane City Nursery
P.O. Box 265 Lane City, TX 77453 409-532-0869

Lillie's Plants
Box 5440 Hwy 59 N Lufkin, TX 75901 409-634-4165

Mark's Nursery
Rt. 1 Box 22A Garwood, TX 77442 409-758-3279

Myra's Garden
Rt. 1 Box 61 E Bay City, TX 77414 409-2543634

Neal's Nursery
P.O. Box 896 Livingston, TX 77350 409-328-0750

New Beginnings Plant Farm
8731 Cedar Dr Lumberton, TX 77657 409-751-4281

Oakwood Orchids
901 Sebesta Rd Angleton, TX 77515 409-848-2721

Old Native State Nursery
270 Pug Road Vidor, TX 77662 409-786-1023

Oyster Creek Garden Center, Inc.
2301 FM 523 Oyster Creek, TX 77541 409-239-2833

Paradise Gardens Inc
5610 Hwy 105 W Conroe, TX 77304 409-756-5572

Paradise Plants Plus
2801 Hwy 69 S Lumberton, TX 77657 409-755-1610

Nursery Specialties Key

Del - Delivery
Bus- Years in operation
% plant stock grown w/in 75 mile radius
Nat - Native plants
Exp - 2 years minimum staff experience
Mem - Member of nursery association
Learn C - Learning center
Fert - Organic fertilizer & pesticide
Demo G - Demonstration garden
News - Newsletter
T Spade - Tree spade
See page 68 for survey details.

Pets & Plants
1230 East Mulberry Angleton, TX 775150000 409-849-0072

Pick-A-Plant
Hwy 96-69 At Cooks Lake Rd. Lumberton, TX 77657 409-755-3331

Piney Woods Nursery
RR 1 BPX 14 Pineland, TX 75968 409-584-2732

Plant It Nursery
3520 W Cardinal Beaumont, TX 77705 409-794-2505
www.plantitnursery.com

Plants N Stuff
201 E Washington Av Navasota, TX 77868 409-825-8477

Pom Pom Nursery and Landscaping
Rt. 2 Box 2532 Brazoria, TX 77422 409-798-5468

Pond's Plus
4626 Hwy 69 S Kountze Beaumont, TX 77705 409-980-3706

Rachuneks Nursery
318 Margaret Street Wharton, TX 77488 409-532-1887

Radke's Nursery
2401 Atkinson Dr Lufkin, TX 75901 409-632-7068

Red Barn Nursery
839 Brooks Rd Wharton, TX 77488 409-532-5140

Rippers Garden Center- C/O Evelyn & D Ripper
2618 F M 1579 Schulenburg, TX 78956 409-561-8420

Shadeview Nursery
4710 North TX Ave. Bryan, TX 77803 409-778-8890

Shell Plant Farm Incorporated
2705 N 10th St Beaumont, TX 77703 409-892-0434

Sky Creek Trees
Route 2 Box 129 East Bernard, TX 77435 409-532-8111

Smiles Garden Center
210 Hickory Creek Dr Shepherd, TX 77371 409-628-3273

Spell Nursery
1200 S 27th St Nederland, TX 77627 409-724-2128

Stewarts Garden Center Inc
194 IH 45 South Huntsville, TX 77340 409-295-5282

Sunshine Herb's And More
1726 21st St. Galveston, TX 77550 409-763-5029
Herbs & herbal vinegars

Sunshine Plant House
3425 Concord Road Beaumont, TX 77703 409-835-7007
Home grown hanging baskets & perennials

Texas Gardens & Landscape
1301 Hwy 290 E Brenham, TX 77833 409-836-4600

The Alamo Nursery
505 Depot St El Campo, TX 77437 409-543-7290

The Cottage Garden Nursery
41 Chapa Rd. (Evergreen) Willis, TX 77378 409-767-4475

The Farm Patch
P.O. Box 4508 | Bryan, TX 77805 | 409-822-7209

The Garden Of Gethsemane
1395 Calvin Dr | Beaumont, TX 77707 | 409-842-2773

The Garden Path
358 N Lhs Dr | Lumberton, TX 77657 | 409-755-2977

The Great Outdoors Company
3602 North St | Nacogdoches, TX 75961 | 409-564-4747

The Plant Environment
1809 Normal Park | Huntsville, TX 77340 | 409-291-0542

The Plant Factory
Rt. 1 Box 322 | Hillister, TX 77624 | 409-547-3636

The West Nursery & Landscape
18399 IH 45 S | Conroe, TX 77385 | 409-273-4040

Timberland Nursery
2202 N. Timberland Dr. | Lufkin, TX 75901 | 409-639-2277

TLC Nursery & Landscape Company
Route 4, Box 1565 | Colmesneil, TX 75938 | 409-837-2852

Tom's Thumb Nursery And Landscaping
2014 45th St. | Galveston, TX 77550 | 409-763-4713

Tropical Gardens Nursery
Rt. 9 Box 163 | Livingston, TX 77351 | 409-327-3529

Vernes Gazebo Gardens
2609 W Hwy 190 | Livingston, TX 77351 | 409-967-4142

Victorian Landscapes Garden And Gifts
4236 S FM 158 | Bryan, TX 77802 | 409-731-8195

Westbury Nursery
403 Gallier Rd. | Beaumont, TX 77713 | 409-860-7201
Del; Bus 5yrs; Nat; Shrubs; Demo G; Everyday 9-6;
www.westburynursery.com

Yucca Do Nursery
P.O. Box 655 | Waller, TX 77484 | 409-826-6363
Rare palms

956 Area Code - Loredo, Brownsville and South Texas

Am-Mex Nursery
914 East Highway 83 | Pharr, TX 78577 | 956-781-2115

Arrangements By Claude's Flower
1920 Convent | Laredo, TX 78040 | 956-724-5913

Barrera's Nursery
P.O. Box 458 | Mc Allen, TX 78505 | 956-686-1706

Basket And Pottery Alley Inc
3519 San Bernardo Ave | Laredo, TX 78041 | 956-726-7671

Brush Country Crafts & Native Plant Nurs
601 E 2nd St | Rio Grande City, TX 78582 | 956-488-2265

Don Juan's Plants
1001 W 2nd | Mercedes, TX 78570 | 956-565-0756

Flores Nursery
P.O. Box 386 | Penitas, TX 78576 | 956-585-4857

G And G Garden Center.- Pablo Gonzalez
4507 Sepulveda Ln. | Laredo, TX 78046 | 956-763-0693

Garden City
P.O. Box 312 | San Benito, TX 78586 | 956-399-3999
Bus.12yrs. 33%; Nat; Demo G; Exp; Fert; Hours M-S 8:30-6,
Sun 12-5

Gentrys Garden Center
4580 North Expressway | Brownsville, TX 78526 | 956-350-9805

Gentry's Laredo Garden Center
P.O. Box 2629 | Laredo, TX 78088 | 956-722-0555

Hollywood Garden Co.
601 Shiloh Dr | Laredo, TX 78045 | 956-724-5769

J O And Y Nursery
1616 Calle De Norte #40 | Laredo, TX 78041 | 956-744-8204

Kays Cactus Gardens
P.O. Box 1240 | Olmito, TX 78575 | 956-350-4032

Los Pinos
4019 Sanders | Laredo, TX 78040 | 956-718-4072

Matt's Garden Center - Charlie Martinez
P.O. Box 322 | Harlingen, TX 78551 | 956-399-3554

Mid Valley Gardens And Ponds
1800 E Hwy 83 | Weslaco, TX 78596 | 956-973-1998

Palm Garden Nursery
345 Galveston Road | Brownsville, TX 78521 | 956-546-1348

Palms And Plants Nursery
Route 1 Box 270d | Donna, TX 78537 | 956-464-4390

Quartermoon Nursery
HC 70 Box 55 | Brownsville, TX 78521 | 956-831-4862

Queen Garden Nursery
4603 San Rafael | Laredo, TX 78043 | 956-712-3732

Rio Gardens
21025 FM 507 | Harlingen, TX 78550 | 956-425-3145

Shary Acres Nursery & Landscaping
3421 N Shary Road | Mission, TX 78572 | 956-581-7783

Southside Nursery
502 Lindenwood | Laredo, TX 78045 | 956-723-0561

Stuart Place Nursery
P.O. Box 2667 | Harlingen, TX 78551 | 956-428-4439

The Oak Tree Co- Raul C Santos Jr.
1718 Reynolds St | Laredo, TX 78043 | 956-725-6697

Things That Grow By Lightouse
1021 Pompano St | Port Isabel, TX 78578 | 956-943-3365

Tonys Nursery
895 East Los Ebanos | Brownsville, TX 78520 | 956-541-5322

Torres Nursery
Rt. 5 Box 8 Teege Rd Harlingen, TX 78552 956-423-6729

Tropical Nursery
Rt. 5 Box 2134 Edinburg, TX 78539 956-383-7234

Upper Valley Fruit Stand And Nursery
Box 2703 Zapata, TX 78076 956-765-3155

Valley Garden Center Inc.
701 E. Hwy 83 McAllen, TX 78501 956-682-9411
Del; Bus. 23 yrs; 33%; Mem; Learn C; Fert; T Spade;
www.valleygardencenter.com

361 Area Code - Corpus Christi, Loredo, Del Rio and Area

Adams Nursery And Flowers
1515 Highway 35 South Rockport, TX 78382 361- 729-7111
Del; Bus.21 yrs; Nat; Large variety of plants; Exp; Mem.;
Learn C; Fert; Hours M-F 9-5, Sat 9-3; Sprg. 7 days a week open.

Annaville Nursery And Florist
11225 Leopard St. Corpus Christi, TX 78410 361 -241-3261

Beale Nursery
P.O. Box 37 Flatonia, TX 78641 361 -865-3967

Better Gardens
606 Mockingbird Victoria, TX 77904 361 -573-7434

Bishop Greenhouses
P.O. Box 363 Driscoll, TX 78351 361 -584-2039

Bobby Lane Nursery
95 Bobby Lane Port Lavaca, TX 77979 361 -552-6297

Bulgerin Brothers Nurseries Inc
Route 4 Box 311 Cuero, TX 77954 361 -275-5683

Country Shoppe Nursery Inc
P.O. Box 1381 Alice, TX 78333 361 -664-8802

Crystal Daisy
2951 Hwy 35 N. Rockport, TX 78362 361 -729-9570

Curry's Nursery And Floral
P.O. Box 813 Three Rivers, TX 78071 361 -786-9921

Double S Nursery
P.O. Box 1731 Beeville, TX 78104 361- 362-0063

Nursery Specialties Key

Del - Delivery
Bus- Years in operation
% plant stock grown w/in 75 mile radius
Nat - Native plants
Exp - 2 years minimum staff experience
Mem - Member of nursery association
Learn C - Learning center
Fert - Organic fertilizer & pesticide
Demo G - Demonstration garden
News - Newsletter
T Spade - Tree spade
See page 68 for survey details.

Earth Work Nursery
102 E Airline Victoria, TX 77901 361-573-3836

Eddie's Nursery
1810 W Market St. Rockport, TX 78382 361-729-3089

El Patio Nursery, Inc.
1704 E. Hillside Laredo, TX 78041 361-724-4558
33%; Nat; Tropicals, plams; Exp; Mem; Fert; T Spade.

Four Seasons Garden Center
1209 East Salem Road Victoria, TX 77901 361-575-8807

Fox Tree And Landscaping Nursery
5949 La Costa Corpus Christi, TX 78413 361-939-7017

Garcias Nursery
1425 Southmost Road Brownsville, TX 78520 361-541-2027

Garden Getaway
4528 CR 56 Robstown, TX 78380 361-387-4350

Gill Landscape Nursery
2810 Airline Corpus Christi, TX 78414 361-993-4795
361-992-9674

Goliad Gardens
510 W Colorado Victoria, TX 77901 361-578-7706

Hillside Nursery
603 West Hillside Laredo, TX 78041 361 724-9988

Janak Brother's Nursery & Landscape
Route 1 Box 126A Shiner, TX 77984 361-594-4321

Janak Nursery Inc
Rt. 1 Box 248 Shiner, TX 77984 361-798-3092

Jo's Green Hut & Nursery
Route 4 Box 249B Hallettsville, TX 77964 361-798-2209

Macaw Village Nursery
15605 Spid Corpus Christi, TX 78418 361-949-7161

Mary's Plants Flowers And Things
5316 FM 624 Robstown, TX 78380 361-387-2732

Mcfarlanes Plants And Garden Supplies
205 Fox Dr Del Rio, TX 78840 361-775-3320

Mendez Nursery #2
Route 5 Box 337 San Benito, TX 78586 361-233-5836

Mini Paradise Nursery
534 Tangelo Edinburg, TX 78539 361-316-2385

P And V Nursery
3218 Mcardle Corpus Christi, TX 78415 361-855-9590

Padre Island Garden Ctr. And Hardware
15925 Spid Corpus Christi, TX 78418 361-949-1963

Padre Palms And Landscape
10015 Spid Corpus Christi, TX 78418 361-937-7700

Petalers Nursery
P.O. Box 519 Sinton, TX 78387 361-364-3531

Renken Nursery Inc
2701 Salem Rd | Victoria, TX 77904 | 361-576-5657

Ressman Tree Service And Nursery
1236 W State Hwy 72 | Yorktown, TX 78164 | 361-564-2248

Saenger's Garden Store
12202 Leopard St. | Corpus Christi, TX 78410 | 361-241-3326

Saenger's Nursery And Landscaping
12201 Leopard St. | Corpus Christi, TX 78410 | 361-241-3326

Sheryl's Country Nursery
8061 FM 1157 | Ganado, TX 77962 | 361-771-2772

Spring Creek Nursery And Gifts
PO Drawer 149 | Freer, TX 78357 | 361-664-4659

Straight Edge Landscape Nursery
P.O. Box 5741 | Kingsville, TX 78364 | 361-592-5939

The Greenhouse
704 North Virginia | Port Lavaca, TX 77979 | 361-552-9758

Tillies Jungle Garden Nursery
P.O. Box 443 | Flatonia, TX 78941 | 361-865-3327

Turner's Nursery And Garden Center
P.O. Box 792 | Premont, TX 78375 | 361-348-3314
Del; Bus 28 yrs; Mem; 33%; www.turnergardenland.com

214 Area Code - Dallas Area

AAA Nursery
10550 Miller Road | Dallas, TX 75238 | 214-342-1794

Cristina's Flowers
6250 Maple Shade Lane | Dallas, TX 75252 | 214-357-5626

Four Seasons Nursery And Landscape
604 N. I-30 Service Rd. | Mount Vernon, TX 75457 | 214-537-4744

Garden View Nursery
7023 Greenville Ave. | Dallas, TX 75231 | 214-373-8890

Gertrude's Garden
5551 W Lovers Lane | Dallas, TX 75209 | 214-357-1804

Hawkins Nursery & Landscaping Co. Inc.
19110 Preston Rd. | Dallas, TX | 972-596-5904
Containerized plants

Into The Garden
5370 W. Lovers Lane | Dallas, TX 75209 | 214-351-5125

LakeJune Garden Center #47
8634 LakeJune Road | Dallas, TX 75217 | 214-391-4005

Lakewood Garden Center Inc #2
6444 E. Mockingbird. Ln | Dallas, TX 75214 | 214-821-9680

Landscape Systems Of Texas
608 S. Pearl st. | Dallas, TX 75201 | 214-748-1435
Trees, shrubs, Tropicals; www.landscapesystems.com

Miller Nursery
10025 Plano Road | Dallas, TX 75238 | 214-348-0810

Nicholson Hardie Nursery
5600 West Lovers Lane, | Dallas, TX 75209 | 214-357-4348

North Dallas Plant Sales
10190 Forest Lane | Dallas, TX 75243 | 214-343-9083

North Haven Gardens
7700 Northaven Rd. | Dallas, TX 75230 | 214-363-5316
Del; Bus.50yrs; 33%; Nat; Perennials, begonias, hostas, roses; Exp; Mem; Demo G; Learn C, News; Fert; Apr-Nov. M-S 9-6, Sun 10-5, Nov-Mar 9-5:30 M-S sun 10-5, Closd Thkg, Xmas, & New Year; www.nhg.com

Once Upon A Thyme
1920 Abrams #273 | Dallas, TX 75214 | 214-828-1020

Orchid Gardens
9748 Brockbank Dr. | Dallas, TX 75220 | 214-350-4985

P and E Nursery and Landscape Supply Inc
1204 S Central | Dallas, TX 75201 | 214-748-7438

Plants And Panters
8550 Plano Road | Dallas, TX 75238 | 214-340-1020

Redenta's Garden
2001 Skillman St. | Dallas, TX | 214-823-9421
Organic garden center

Ruibal's Plants Of Texas Inc
1118 S Central | Dallas, TX 75201 | 214-744-9991

Strong's Nursery & Garden Center
1900 N. IH 35E | Carrollton, TX 75006 | 972-242-7574
Flowering plants

Sunshine Miniature Trees
7118 Greenville | Dallas, TX 75231 | 214-691-0127
Bonsais & houseplants

Texas Botanical Garden
8522 Garland Road | Dallas, TX 75218 | 214-328-6806

Texas Palm Trees And Plants
2023 Cadiz | Dallas, TX 75201 | 214-741-2310

Urban Gardener Inc
305 Centre St. | Dallas, TX 75208 | 214-943-6785

W W Tree Farms
11814 Harry Hines Blvd | Dallas, TX | 214-748-1435
Treesn shrubs, tropicals, Large Trees.

972 Area Code - Dallas and Ft. Worth Area

9T Farms Nursery
P.O. Box 640 | Ennis, TX 75120 | 972-875-8369

Big John's Nursery And Landscaping Co.
621 Straus Rd. | Cedar Hill, TX 75105 | 972-291-2231

Black Eyed Susan's Inc
810 Parker Square | Flower Mound, TX 75028 | 972-899-8888

Blooming Color Nursery & Lndscping Inc
1701 E. Beltline Rd. | Coppell, TX 75019 | 972-393-8660
Del, Bus.10yrs; 33%; Nat; Perennials, Native trees, Specimen trees; Exp; Mem; Demo G; Learn C; Fert; www.bloomingcolorsnursery.com

Bluebonnet Garden Center
530 E Hwy 287 Midlothian, TX 76065 972-775-4434

Bluebonnet Nursery
2355 West FM 544 Wylie, TX 75098 972-442-2734

Bruce Miller's Nursery
1000 E. Belt Line Rd. Richardson, TX 75081 972-238-0204

Bundy Nursery
P.O. Box 807 McKinney, TX 75069 972-542-4052

Calloway Nurseries
4200 Airport Frwy. #200 Fort Worth, TX 76117 972-964-3084
1621 Custer Rd. Plano, TX 972-596-5211
2100 N. Plano Rd. Richardson, TX 972-644-0144
www.calloways.com

Central Garden And Pet
P.O. Box 655650 Dallas, TX 75265 972-466-0069

Chambers Nursery And Garden Center
3307 W Shady Grove Irving, TX 75060 972-986-5903

Cheakas Home And Garden
400 Boyd St. Allen, TX 75002 972-359-0984

Classic Gardens And Landscape Company
401 W FM 544 Murphy, TX 75094 972-424-9929

Cottonwood Creek Nursery
Rt. 3, Box 4 Frisco, TX 75034 972-335-2988

Covington Landscape And Nursery Inc
P.O. Box 1165 Rowlett, TX 75030 972-475-5888

Creative Water Gardens
2125 W. Kingsley Garland, TX 75041 972-271-1411

Doan's Nursery
622 S. Beltline Dallas, TX 75253 972-790-3500

Foster Greenhouse And Nursery
506 E Mulberry Kaufman, TX 75142 972-932-6441

Foster's Garden Center
8112 S. Hampton Dallas, TX 75232 972-224-0131

Fosters Garden Center
2305 W. Wheatland Rd Dallas, TX 75232 972-224-0131

Garrison Gardens Incorporated
1817 Avenue K Plano, TX 75074 972-424-7568

Nursery Specialties Key

Del - Delivery
Bus- Years in operation
% plant stock grown w/in 75 mile radius
Nat - Native plants
Exp - 2 years minimum staff experience
Mem - Member of nursery association
Learn C - Learning center
Fert - Organic fertilizer & pesticide
Demo G - Demonstration garden
News - Newsletter
T Spade - Tree spade
See page 68 for survey details.

Gibson Discount Garden Center
P.O. Box 154392 Irving, TX 75015 972-254-3989

Greenhouse Gardens- North Texas Landscape Co.
4402 W. University McKinney, TX 75070 972-562-5895

Gunter's Greenhouse Inc
513 W. Campbell Rd Richardson, TX 75080 972-234-6017
Del; Bus 29yrs; Worldwide variety of flowers; www.gunters.com

Hartwells Nursery
1570 N. Stemmons Lewisville, TX 75067 972-436-3612

Hawkins Nursery And Landscape
19110 Preston Rd. Dallas, TX 75252 972-596-5904

Hogan Nursery And Landscape Supply
2409-Erwin Drive Plano, TX 75074 972-424-4662

Homesley Nursery
707 W Broad St Forney, TX 75126 972-564-1426

Huggins Nursery
P.O. Box 294978 Lewisville, TX 75029 972-539-4011

Huggins Nursery & Plant Center
1616 Arrowhead Dr. Flower Mound, TX 75028 972-539-4011

J And T Nursery And Landscaping
5126 W US Hwy 80 Terrell, TX 75160 972-524-0806

Lawsons Nursery And Landscape
1601 West Moore Terrell, TX 75160 972-563-4381

Lindsey's Lnadscape Design Center
3502 Hwy 77 N Bldg B Waxahachie, TX 75165 972-617-9717

Miller Nursery And Landscape
1000 East Belt Line Richardson, TX 75081 972-238-0204

Monday Landscape Nursery Inc
4022 Broadway Garland, TX 75043 972-240-2606

North LakeNursery And Landscaping Co.
794 Denton Tap Coppell, TX 75019 972-462-1916

Oak Tree Nursery Inc.
705 Highway 175 Seagoville, TX 75159 972-287-4504

Pat's Plants #2
529 South Carrier Grand Prairie, TX 75051 972-264-7487

Plants And Planters
1050 North Greenville Ave. Richardson, TX 75081 972-699-1281

Rainbow Gardens Nursery
P.O. Box 767 3 Stores Allen, TX 75013 972-727-9338

Raymundo Nursery & Landscaping
860 South Highway 5 Fairview, TX 75069 972-542-7812

Rohde's Nursery
1651 Wall St. Dallas, TX 75215 972-864-1934

S and P Home and Garden Treasures
1353 Jasmine Dr. Lewisville, TX 75077 972-317-7004

Shades Of Green Incorporated
5400 B Preston Rd. | Frisco, TX 75034 | 972-335-9095

Southwest Landscape Nursery Company
2220 Sandy Lake Rd. | Carrollton, TX 75006 | 972-245-4557

Strong's Nursery
1900 N I-35 E | Carrollton, TX 75006 | 972-242-7574

Sunday Nursery and Landscape
2049-A S. Belt Line Rd. | Dallas, TX 75253 | 972-285-4122

Texas Seasons Corp
6447 Preston Rd. | Frisco, TX 75034 | 972-335-9446

The Greenery
3708 N. Hwy 77 | Waxahachie, TX 75165 | 972-617-5459

The Greenstem Inc.
3950 Clear Cove | Dallas, TX 75244 | 972-243-3417

The Plaid Veranda Inc
7127 Main St | Frisco, TX 75034 | 972-377-8080

The Plant Market Inc
500 Spring Creek Village | Dallas, TX 75248 | 972-239-0049

The Plant Shed
740 Lexington Dr. | Plano, TX 75075 | 972-516-2298

The Rose Petal
515 E Main | Grand Prairie, TX 75050 | 972-262-5253

Vaughans Backyard Garden Center
804 Ferris Ave | Waxahachie, TX 75165 | 972-937-2531

Village Plant Farm Inc
P.O. Box 319 | Rosser, TX 75157 | 972-486-3200

915 Area Code - El Paso, Abilene, Midland, Odessa, San Angelo and Area

Acco Farm & Ranch Center
249 South 11th | Abilene, TX 79602 | 915-676-3302

Agua Etc.
P.O. Box 497 | McCamey, TX 79752 | 915-652-3453

Alldredge Gardens
P.O. Box 50005 | Midland, TX 79710 | 915-682-4500

Baacks Inc
1842 Matador | Abilene, TX 79605 | 915-692-7812

Baacks Incorporated
5001 Buffalo Gap | Abilene, TX 79605 | 915-692-7763

Bangs Greenhouse
400 E Hall | Bangs, TX 76823 | 915-752-0212

Beto's Nursery
1217 Westway | Canutillo, TX 79835 | 915-886-5039

Bryan's Nursery
1400 Johnson | Pecos, TX 79772 | 915-447-9037

Caby's Landscape Garden Center Inc.
4601 Hondo Pass Av. | El Paso, TX 79904 | 915-755-5661

Carrillos Nursery
7546 Doniphan Dr. | El Paso, TX 79932 | 915-377-5696

Casa Verde Nursery
77 Fountain | El Paso, TX 79912 | 915-584-1149

City Creek Farms
13395 Hwy 8384 | Tuscola, TX 73062 | 915-670-7434

Country Gardens- James Richardson
1925 Ford Street | Llano, TX 78643 | 915-247-3387

Decker's Farm Supply & Nursery
P.O. Box 1491 | Stanton, TX 79782 | 915-756-3444

Desertland Nursery & Pottery Center
11306 Gateway Blvd. East | El Paso, TX 79927 | 915-858-1130
Bus. 33yrs; 33%; Nat; Exotic Cacti; Exp; Mem; hours 8-6 M-Sat, 10-6 Sun

Earlene's Florist And Landscaping
P.O. Box 731 | Goldthwaite, TX 76844 | 915-648-2478

Garden Mart Incorporated, The
4012 North Mesa | El Paso, TX 79902 | 915-533-6524

Garden Place
4002 North 1st | Abilene, TX 79603 | 915-676-0086

Garden World- Larry Brewster Inc
2850 S Clack | Abilene, TX 79606 | 915-698-2401

Gardens Of The Southwest Inc
5250 South 14th Street | Abilene, TX 79605 | 915-692-1457

Gonxales Cacti Nursery
5627 Charolais | El Paso, TX | 505-382-7272
800-330-5821

Gwen's Greehouse
129 N Hwy 84 | Snyder, TX 79549 | 915-573-5153

H And H Nursery
218 North Access Rd. West | Clyde, TX 79510 | 915-893-2069

Hepburns Plantland & Florist
HC 70 Box 3 | Brady, TX 76825 | 915-597-0736

Horton Hill Nursery
P.O. Box 133 | Roby, TX 79543 | 915-776-2809

JC's Tree Farm
2122 S. Eddy | Pecos, TX 79772 | 915-447-9701

Kalico West Nursery
100 W Waco | Kermit, TX 79745 | 915-586-6730

LaCasa Verde Nursery Inc
2615 N Midland Dr | Midland, TX 79707 | 915-520-2144

LaMaceta Garden Center
3621 Lee Blvd. | El Paso, TX 79936 | 915-855-4444

Lemon Street Nursery
1301 Lemon | Sweetwater, TX 79556 | 915-235-4459

Limbs And Leaves Nursery
Hc07 Box 126 A | Llano, TX 78643 | 915-247-3323

M And R Nurseries Incorporated
1601 West Beauregard | San Angelo, TX 76901 | 915-653-3341

Magic Farms Greenhouses
7678 Country Club Road | San Angelo, TX 76904 | 915-944-0617
Del; Bus.18yrs; Geraniums, Groundcovers, Pointsettias; Exp 2 yrs;
Open 7 days/week, hours 8-7

Mannings Garden Center
2820 Golf Course Road | Midland, TX 79701 | 915-682-8533

Nancy's Nursery Plants And Things Inc
11355 Pellicano Dr. | El Paso, TX 79936 | 915-598-3435

Native Ornamentals
110 S 6th Street | Mertzn, TX 76941 | 915-835-2021

Neffs Nursery
P.O. Box 1309 | Colorado City, TX 79512 | 915-728-2428

Odessa Nursery Inc.
4306 Andrews Highway | Odessa, TX 79762 | 915-366-6781

Olive Nursery Incorporated
3402 Sherwood Way | San Angelo, TX 76901 | 915-949-3756

Palmas Nursery & Landscaping
1111 Coe Battle Blvd. | El Paso, TX 79927 | 915-860-1900

Payless Nursery Inc.
8000 Gateway East | El Paso, TX 79907 | 915-591-9337

Pearsons Tree Place
6900 Doniphan | Canutillo, TX 79835 | 915-877-3808

Plantation Discount Nursery
105 Lindberg Ave. | El Paso, TX 79932 | 915-584-0433

Plantation Nursery
5372 Doniphan Dr. | El Paso, TX 79932 | 915-585-1131

Plants Galore And More
807 Commercial | Anson, TX 79501 | 915-823-4660

Ponderosa Nursery
204 Davis Rd | Big Spring, TX 79720 | 915-263-0260

Ricky Enterprise Nursery- Ricky Navar
1773 Pali Dr. | El Paso, TX 79936 | 915-855-7474

Nursery Specialties Key

Del - Delivery
Bus- Years in operation
% plant stock grown w/in 75 mile radius
Nat - Native plants
Exp - 2 years minimum staff experience
Mem - Member of nursery association
Learn C - Learning center
Fert - Organic fertilizer & pesticide
Demo G - Demonstration garden
News - Newsletter
T Spade - Tree spade

See page 68 for survey details.

Roberts Greenhouse
1104 Cardinal Crest | Andrews, TX 79714 | 915-524-3265

Scherz Landscape Co.
2225 Knickerbocker Rd. | San Angelo, TX 76904 | 915-944-0511

Scott's Nursery
4044 S Bryant Blvd | San Angelo, TX 76903 | 915-653-7723

Sunshine Garden And Gifts Inc
5253 W University | Odessa, TX 79764 | 915-381-1888

The Plant And Garden Place
4577 IH 20 W | Baird, TX 79504 | 915-893-4829

The Plantery
315 N. Pope | San Angelo, TX 76903 | 915-653-1496

The Rustic Wheelbarrow
416 W. Ave. D | San Angelo, TX 76903 | 915-659-2130

The Secrect Garden
405 W. Lake Dr. | Hamlin, TX 79520 | 915-576-2538

Treehouse Nursery
P.O. Box 261 | Seminole, TX 79360 | 915-758-0007

505 Area Code - New Mexico

Gonzales Cacti Nursery
5627 Charolais Dr. | Las Cruces, NM 88012 | 505-382-7272

Sierra Vistat Growers
2800 Nm Hwy 28 | Las Cruces, NM 88005 | 505-874-2415

817 Area Code - Ft. Worth and Area

Acorn Garden Center
701 Old Brock Rd | Weatherford, TX 76088 | 817 -596-4117

Armadillo Nursery
1105 S Caddo | Cleburne, TX 76031 | 817 -517-7337

Artistic Plants
608 Holly Drive | Burleson, TX 76028 | 817 -295-0802

Bean's Home And Garden
533 W Bolt St | Fort Worth, TX 76110 | 817 -923-6932

Beautiful Things
3733 Bellaire Dr. South | Fort Worth, TX 76109 | 817 -924-1935

Blooming Color Nursery & Landscape
221 IRA Woods (Hwy 26) | Grapevine, TX 76051 | 817-416-6669
Del, Bus.10yrs., 33%; Nat; Perennials; Exp; Mem.; Demo G; News;
Learn C; Fert; Hours M-S 9-6, Sun 12-5 www.bloomingcolorsnursery.com

Busby Nursery
7925 Jacksboro Hwy | Fort Worth, TX 76135 | 817 -237-4884

Butterflies Over Texas
2510 Riverview Trail | Granbury, TX 76048 | 817 -576-2919
Bus. 1yr; 33%; Nat; Butterfly Plants; Demo G; Learn C; Fert;
www.butterfliesovertexas.com

Casa Verde Nursery & Landscape
2720 Horne St | Fort Worth, TX 76107 | 817 -737-7843

CC's Touch Of Nature
3912 West Vickery — Fort Worth, TX 76107 — 817- 732-0942

Calloway Nursery
2651 S. Hulen — Ft. Worth, TX — 817-923-9979
3525 Hwy 121 — So. Grapevine, TX — 817-421-6667
760 Grapevine Hwy — Hurst, TX — 817-581-6361

Cole's Plants Inc.
5807 Quanah Hill Rd. — Dallas, TX 75228 — 817-498-5033

Cooper Street Farmers Market
1606 South Cooper — Arlington, TX 76013 — 817 -276-8810

Cornelius
1755 W. FM Rd. 1960 — Houston, TX 77057 — 817- 222-1122

Cox's Plant Farm
2405 W. Harris Rd. — Arlington, TX 76001 — 817-467-2431

Farmers Market Ft Worth Inc
5507 East Belknap — Haltom City, TX 76117 — 817 -838-8526

Flora's Flowers And Plants
713-Cr 702 — Cleburne, TX 76031 — 817 -645-2308

Foltz And Son
1204 Hwy 156 S. — Haslet, TX 76052 — 817-439-0055

Glade Road Garden Center
1221 West Glade Road — Euless, TX 76039 — 817 -283-2781

Granbury Nursery
2410 Hwy 377 E — Granbury, TX 76048 — 817 -573-1251
Del; Bus.33yrs; Exp; LernC; Fert; Hours M-S 8-5:30, Sun open in season

Green Mama's
5324 Davis Blvd — N. Richland Hills, TX 76180 — 817 -514-7336

Guerin Nursery
Hwy 167 South Acton — Grandbury, TX 76049 — 817-326-2507

Hamilton Nursery
P.O. Box 501 — Groesbeck, TX 76642 — 817 -729-2407

Hare's Nursery
3200 W Division — Arlington, TX 76012 — 817-275-1261

Hilscher Nursery & Garden Center
5015 East Lancaster — Fort Worth, TX 76103 — 817 -536-8369

Home To Garden
3986 W Vickery — Fort Worth, TX 76107 — 817 -377-0318

In The Garden
1411 N Main St — Cleburne, TX 76031 — 817- 774-9000

Into The Garden
1612 University — Fort Worth, TX 76107 — 817-336-4686

Jackalope Of Texas
5925 Camp Bowie — Fort Worth, TX 76107 — 817- 737-7687

Lee's Nursery
7803 S. Cooper street — Arlington, TX 76001 — 817 -465-0065
Del; Bus 3yrs; 33%; Nat; Larger trees; Exp; News; Hours 12-4 M-Sat.

M J Farm & Nursery
2499 Union Church Rd — Keller, TX 76248 — 817-431-5915

Mansfield Nursery
1213 Glen Creek Drive — Mansfield, TX 76063 — 817 -473-1514

Marshall Grain Co Inc
2224 E Lancaster — Ft Worth, TX 76103 — 817-536-5636

Metro Maples
4890 S Dick Price Rd. — Fort Worth, TX 76140 — 817-797-3419

Mike's Garden Center Inc- Mike Cook
5703 Crowley Road — Fort Worth, TX 76134 — 817- 293-8800

Nature's Treasures
1500 Florence — Azle, TX 76020 — 817- 237-0785

Plant Peddler
1115 Walters Dr. — Granbury, TX 76048 — 817- 279-1230

Plant Shed Inc
1501a W Airport Frwy — Euless, TX 76040 — 817 -346-8700

Plants Of Fort Worth
4802 Hwy 377 S — Fort Worth, TX 76116 — 817- 249-1315

Plantus Nursery
7729 Highway 80 West — Fort Worth, TX 76116 — 817 -244-2250

Rainbow Plant Sales
6115 Garcia Ln. — Ft Worth, TX 76140 — 817- 478-1051

Redenta's Garden
5111 W Arkansas Ln. — Arlington, TX 76016 — 817- 451-2149

Robert's Native Trees
805 Mc Anear — Cleburne, TX 76031 — 817- 641-9435

Stuart Nursery
2317 Ft Worth Hwy. — Weatherford, TX 76087 — 817-596-0003
Bus.30 yrs; 33%; Nat; Demo G; Exp; Mem; Fert.

Tarver Nursery
313 Byron — Fort Worth, TX 76114 — 817-732-2782

The Plant Shed
1501a W Airport Fwy. — Euless, TX 76040 — 817- 540-4580

The Rock Garden
7844 Grassland Dr — Fort Worth, TX 76133 — 817 -346-3792

Therrell's Family Garden Center
2505 Hemphill St. — Fort Worth, TX 76110 — 817-924-3903

Tropical Greenery
7908 NE Loop 820 — Fort Worth, TX 76180 — 817- 485-2100

Weston Gardens In Bloom Inc.
8101 Anglin Dr. — Fort Worth, TX 76140 — 817-572-0549
Perennials, antique roses, ornamental grasses; Nat; Demo G; Fert;
www.westongardens.com

Willow LakeFarm & Nursery
3400 Mineral Wells Hwy. — Weatherford, TX 76087 — 817-613-1486

713 Area Code - Downtown Houston Area

Allure
5403 S Rice | Houston, TX 77081 | 713-669-8188

Anderson Landscape & Nursery, Inc.
2222 Pech Road | Houston, TX 77055 | 713-984-1342

Another Place In Time
1102 Tulane Street | Houston, TX 77008 | 713-864-9717

B & L Nursery
11519 N. Garden | Houston, TX 77071 | 713-772-9917

Bella Floral LLC
111 Heights Blvd | Houston, TX 77007 | 713-426-5556

Brazos Oaks Nursery
9999 Southwest Freeway | Houston, TX 77074 | 713-771-6257

Brookston Nursery
5207 Brookston | Houston, TX 77045 | 713-729-6627

Brookwood Garden Shop
1752 FM 1489 | Brookshire, TX 77423 | 713-782-7757

Buchanans Native Plants
611 E. 11th St. | Houston, TX 77008 | 713-861-5702

Candys Garden Center
5417 Anderson Rd At Post Oak | Houston, TX 77053 | 713-434-0552

Condon Gardens Incorporated
1214 Augusta Dr | Houston, TX 77057 | 713-782-3992
Unusual

Congo Nursery
2304 Wichita #125 | Pasadena, TX 77502 | 713-475-1081

Cornelius Nurseries Inc.
2233 South Voss Rd | Houston, TX 77057 | 713-782-7611
Shrubs, silk & dried flowers

Cornelius
1200 Ashford | Houston, TX 77057 | 713-782-8640

Coulson Plants
12665 Coulson | Houston, TX 77015 | 713-455-3380

Donaho Garden Center Inc
3402 Fondren | Houston, TX 77063 | 713-781-0702

Flamingo Gardens, Inc.
9600 Old Katy Road | Houston, TX 77055 | 713-468-5600

Nursery Specialties Key

Del - Delivery
Bus- Years in operation
% plant stock grown w/in 75 mile radius
Nat - Native plants
Exp - 2 years minimum staff experience
Mem - Member of nursery association
Learn C - Learning center
Fert - Organic fertilizer & pesticide
Demo G - Demonstration garden
News - Newsletter
T Spade - Tree spade
See page 68 for survey details.

Four O Clock Plant Nursery
302 Rustic Dr | Pasadena, TX 77502 | 713-475-1341

Gallery Flowers And Garden Ctr.
4002 Fannin | Houston, TX 77004 | 713-522-6692

Gateway Ace Hardware & Garden Center Inc
6860 Telephone | Houston, TX 77061 | 713-643-0623

Glauser-Mcnair, Inc.
1707 Ojeman | Houston, TX 77055 | 713-781-3841

Green Acres Nursery
1822 Tidwell Lane | Houston, TX 77093 | 713-694-6781

Houston Flowery
5930 Westheimer | Houston, TX 77057 | 713-977-5350

Houston Garden Center
5000 SW Freeway Parkway N | Houston, TX 77056 | 713-626-1849
2811 Airline Dr. | Houston, TX | 713-869-9505
2732 W. 18th | Houston, TX | 713-869-8870

Joseph's Nursery
8150 SW Frwy @ Beechnut | Houston, TX | 713-772-8002
Large selection

Joshua's Native Plants & Garden Antiques
111 Heights Blvd. | Houston, TX 77007 | 713-862-7444

Lees Hardware & Nursery Incorporated
3620 Tidwell | Houston, TX 77093 | 713-697-6110

Lilypon's Water Gardens
839 FM 1489 | Brookshire, TX 77432 | 1-800-723-7667

Lowe's Watergarden And Etc
17122 Stuebner Airline | Spring, TX 77379 | 713-271-1707

Moellers Greenhouse
1116 Baumgarten St | Schulenburg, TX 78956 | 713-743-3430

Orchids And Ferns
7802 Bellaire Blvd. | Houston, TX 77036 | 713-774-0949

Plant House
812 Westheimer | Houston, TX 77006 | 713-529-6050

Plants N Petals Inc
3810 Westheimer | Houston, TX 77027 | 713-840-9191

Rentokil Tropical Plants
6205 Skyline Dr | Houston, TX 77057 | 713-783-4500

River Gardens
7215 Ashburn | Houston, TX 77061 | 713-645-5988

Robert Taylor Nursery
8809 Homestead | Houston, TX 77016 | 713-635-8525

Roberto's Evergreens
4302 Ireland Street | Houston, TX 77093 | 713-697-2815

Robertsons Nursery
2905 Pasadena Blvd. | Pasadena, TX 77503 | 713-473-1333

Seven Oaks Nursery
10801 Telephone Road | Houston, TX 77075 | 713-991-3330

Tall Plants
14603 Tab Lane | Houston, TX 77070 | 713-464-8671

Teas Nursery Co Inc
4400 Bellaire Blvd., | Bellaire, TX 77401 | 713-664-4400
4545 Beechnut | Meyerland, TX | 713-665-6852
Del; Bus.158 yrs; 33%; Nat; Roses, Azaleas, Crape Myrtle; Demo G; Exp 2 yrs; Mem; Learn C; News, Fert; TSpade; www.teasnursery.com

Thompson-Hanson Nursery , Inc.
2714 Joanel | Houston, TX 77027 | 713-622-3722

Touch Of Green
6747 Signat Dr. | Houston, TX 77041 | 713-861-6918

Trinity Valley Nursery
Route 3 Box 203C | Dayton, TX 77535 | 713-258-7436

US Trees Incorporated
4113 Willowbend | Houston, TX 77025 | 713-660-9223

281 Area Code - South West Houston Area

A Farms
25615 Southwest Frwy. | Rosenberg, TX 77471 | 281-342-0123

Abilas Nursery
16106 Old Richmond Rd | Sugar Land, TX 77478 | 281-495-5342

All Seasons Garden Center
205 N Friendswood Dr | Friendswood, TX 77546 | 281-482-2850

All Seasons Nursery
312 John Alber | Houston, TX 77076 | 281-591-7630

All Seasons Nursery
4901 Vista | Pasadena, TX 77505 | 281-487-8733

All Star Landscaping And Nursery
P.O. Box 597 | Magnolia, TX 77355 | 281-356-3870

Allied Ace Hardware And Garden Center
5203 Fairmont Pkwy | Pasadena, TX 77505 | 281-487-9288

Alpine Trees And Shrubs
21828 Franz | Katy, TX 77449 | 281-578-0627

Alvin Garden Center
1201 Booth Lane | Alvin, TX 77511 | 281-585-2700

Asia Ez Bonsai Nursery Inc
6706 Broadway | Pearland, TX 77581 | 281-997-2835

Azalea Max Nursery
5207 Melanie | Crosby, TX 77532 | 281-328-8736

Barbers Hill Nursery And Landscaping
2325 N FM 565 N | Mont Belvieu, TX 77520 | 281-385-2169

Bev's Trees And More Inc
14607 1/2 Boudreaux | Tomball, TX 77375 | 281-351-0281

Bill Bownd's Nursery
10519 FM 1464 | Richmond, TX 77469 | 281-277-2033

Bownds Select Trees
10519 FM 1464 | Richmond, TX 77469 | 281-277-2033

Budget Nursery
306 Maplewood Dr | Friendswood, TX 77546 | 281-482-0132

Buds And Blossoms
14120 Cypress N. Houston Rd. | Houston, TX 77065 | 281-469-3378

Buchanan's Native Plants
611 E. 11th Street | Houston, TX | 281-861-5702

Butler's Pond And Garden
P.O. Box 356 | League City, TX 77574 | 281-996-0557
Water garden plants; please call ahead for hours; www.pondgirl.com

Cabrera Farm Nursery
P.O. Box 16369 | Sugarland, TX 77496 | 281-980-4667

Cattails Pond And Garden Supply
311 Edgewood Ave. | Houston, TX 77011 | 281-482-0840

Champion Top Soil, Inc.
1463 Highway 6 South | Houston, TX 77077 | 281-558-9948

Chenango Gardens And Florist
Rt. 2 Box 1820 Cr 36 | Angleton, TX 77515 | 281-369-2392

Cliftons Country Gardens
P.O. Box 2171 | Crosby, TX 77532 | 281-328-1209

Coco's Nursery Inc
474 Buoy Rd | Webster, TX 77598 | 281-480-9179

Cornelius Nursery Incorporated
P.O. Box 570485 | Houston, TX 77257 | 281-444-1210

Cypress Gardens
3119 Lily | Pasadena, TX 77505 | 281-487-0185

Discount Nursery Inc
802 Hardesty | Seabrook, TX 77586 | 281-474-2417

Discount Waterlilies And Pond Supplies
Rural Route 1 Box 47-T | Brookshire, TX 77423 | 281-375-2495

DJ's Plants And More
504 FM 517 W | Dickinson, TX 77539 | 281-534-4172

Donaho Garden Center Inc
2920 Dairy Ashford | Houston, TX 77082 | 281-589-0037

Enchanted Gardens
6420 FM 359, P.O. Box 280 | Richmond, TX 74406 | 281-341-1206

Flamingo Gardens
11600 Louetta Rd | Houston, TX 77070 | 281-655-0058

Flower And Plants Of Heaven Co.
1582 Tallshadows Dr | Houston, TX 77032 | 281-442-0528

Frank's N Nursery
800 Hwy 90 | Richmond, TX 77469 | 281-342-3211

Fred Reyes Nursery Inc
P.O. Box 519 | Simonton, TX 77476 | 281-346-1442

Friendswood Garden Center Lee's Landscape Co.
907 S. Friendswood Dr | Friendswood, TX 77546 | 281-482-7322

Frio Canyon Nurseries
506 Bayou Dr. Houston, TX 77020 281-351-4804

Front Porch Nursery & Mulch
21411 FM 2920 Hockley, TX 77447 281-351-7130

Garden Dreams
13424-B Briar Forest Houston, TX 77077 281-558-1112
Del; Bus 3yrs; 33%; TX natives, perennials; Exp; Mem; Demo G; News; Fert; www.texasgardendreams.com

Garden Ville Square
2919 N Main Stafford, TX 77477 281-499-6995

Garden Works By Kim
9605 Rustic Gate Laporte, TX 77571 281-542-9004

Gordon Atkins Nursery
P.O. Box 841677 Houston, TX 77284 281-550-6898

Great Western Growers
10655 Stancliff Houston, TX 77099 281-530-0550

Houston Bonsai Farm
1219 TX Pkwy Missouri City, TX 77489 281-261-6819

Houston Garden Center
919 W. Sam Houston Prkwy N Houston, TX 713-973-1478

Houston Palm Tree Inc
20420 Gulf Freeway Webster, TX 77598 281-338-2658

Houston Plants And Garden
16726 North Frwy. Houston, TX 77090 281-358-7704

Houston Plants Garden World Inc
16726 North Frwy. Houston, TX 77090 281-443-3188

Huffman Nursery And Farmers Market
12110 FM 1960 E Huffman, TX 77336 281-324-2911

Huntsville Gardens Inc.
2201 Connorvale Houston, TX 77039 281-449-2258

John's Trees
3902 FM 2920 Spring, TX 77388 281-353-9540

Joseph Nursery
3723 Manvel Rd. Rearland, TX 77584 281-489-9786

JRN Nursery
4809 Breen Dr. Houston, TX 77086 281-448-6107

Nursery Specialties Key

Del - Delivery
Bus- Years in operation
% plant stock grown w/in 75 mile radius
Nat - Native plants
Exp - 2 years minimum staff experience
Mem - Member of nursery association
Learn C - Learning center
Fert - Organic fertilizer & pesticide
Demo G - Demonstration garden
News - Newsletter
T Spade - Tree spade
See page 68 for survey details.

Kingwood Farmers Market
22310 Loop 494 Kingwood, TX 77339 281-359-3765

Kingwood Garden Center Incorporated
1216 Stonehollow Kingwood, TX 77339 281-358-1805

Maas Second Street Nursery
5511 Todville Seabrook, TX 77586 281-474-4974

MB Nursery Inc.
10680 Jones Rd. Houston, TX 77065 281-894-1128

Nature's Touch Garden Ctr.
5165 Louetta Spring, TX 77379 281-655-7171

Nelson Water Gardens & Nursery Inc
P.O. Box 181 Pattison, TX 77466 281-375-8356

Nelson Water Gardens And Nursery Inc
1502 Katy Fort Bend Cr. Katy, TX 77493 281-391-4769

Newport Nursery And Landscape
P.O. Box 1245 Crosby, TX 77532 281-328-1178

Northwest Tree Farms
25819 W. Memory Ln. Houston, TX 77355 281-356-1773

Oasis Palm Tree Co
217 Spencer Landing W La Porte, TX 77571 281-334-9125

Olympia Orchids
27A Windsong Lane Friendswood, TX 77564
Bus 19yrs; Orchids; www.olympiaorchids.com

Payne's in the Grass Daylilly Farm
Pearland, TX 281-485-3821
Daylillies

Pecan Hill Nursery
34303 Pecan Hill Dr. Brookshire, TX 77432 281-346-2001

Plant Cheap Nursery
15925 First St Splendora, TX 77372 281-974-8354

Plant Menagerie
3810 FM 359 Richmond, TX 77469 281-342-6217

Plants And Plantings
12407 Eiker Pearland, TX 775812650 281-485-5954

Plants For All Seasons
6610 Lovetta Rd. Spring, TX 77379 281-370-6390

Plants For All Seasons Nursery
21328 Hwy 249 Houston, TX 77070 281-376-1646

Red Barn Nursery
13105 CR 185 Alvin, TX 77511 281-331-6652

Rosehill Herbs
14914 Treichel Tomball, TX 77375 281-351-2641

Roseland Flower And Nursery Inc.
4824A West Broadway Pealan, TX 77584 281-997-7554

Ryan's Nursery
2116 Fry Road Houston, TX 77084 281-492-2346

Sharon's Garden Center
11907 Hedgegate Dr Houston, TX 77065 281-890-8520

Silver Leaf Nursery
Bill Cornelius 5750 Shirley Ln Houston, TX 77032 281-449-0413

Spring Gardens Nursery
18811 Tomato Spring, TX 77379 281-379-2901

Sun Land Nursery
1901 Cripple Creek Pearland, TX 77581 281-332-5968

Teas Nursery Co. Inc.
25598 I-45 N Woodlands, TX 77401 281-367-2013,
1445 W. Bay Area Blvd. Clear Lake, TX 77401 281-338-2050,
3600 S.E. Beltway Bellaire, TX 77401 281-978-3454,
Del; Bus.158 yrs; 33%; Nat; Roses, Azaleas, Crape Myrtle; Demo G; Exp; Mem; Learn C; News, Fert; TSpade; www.teasnursery.com

Texas Plant And Garden
23598 Loop 494 Porter, TX 77365 281-354-4595

Texas Trees
P.O. Box 5666 Kingwood, TX 77325 281-354-6111

The Savage Garden
907 Taylor St Houston, TX 77007 281-682-7263

The Woodlands Gardens
703 Sawdust Road Spring, TX 77380 281-364-9769

The Yard Sale
Ernestine L Cain 316 Hubert Cleveland, TX 77327 281-592-8319

Theiss Nursery
2192 FM 1488 W Magnolia, TX 77355 281-259-4028

Trees And Plants Incorporated
16205 Capri Houston, TX 77040 281-955-0156

Tropical Garden Scenes
5990 N Sam Houston Pky #602 Humble, TX 77396 281-441-3739

Udemis Nursery
P.O. Box 566 Pattison, TX 77466 281-375-8800

Uncle Ed's Nursery
33130 Hwy 249 Pinehurst, TX 77362 281-356-4200

US Bonsai And Garden Stone
14203 Stuebner Airline Dr. Houston, TX 77069 281-880-7879

Vaughans Nursery And Gardens
15647 Avenue C Channelview, TX 77530 281-452-7369
Del; Bus.25yrs; 33%; Nat; Unusual and hard to find; Mem; Demo G; Learn C; Fert; Open 8-6 seven days a wk. Closed NewYear and Thnksg.
E-mail: maria@vaughnsnursery.com

231 Area Code - Seabrook

Maas Nursery & Landscaping
5511 Todville Rd. Seabrook, TX 77586 231-474-2488

210 Area Code - San Antonio

A and M Nursery
1502 E. Ashley Rd. San Antonio, TX 78214 210- 927-3177

Alfaro Tree Sales Inc.
10835 Old Pearsol Rd. Atosa, TX 78002 210-622-9875

Aztec Gardens
4459 Pleasanton Rd. San Antonio, TX 78221 210-924-0204

Carrasco's Landscape & Nursery
5247 Pearsall Rd. San Antonio, TX 78242 210-623-1276

Castle Hills Nursery
6804 West Ave. San Antonio, TX 78213 210- 349-6461

D And R Landscaping Tree Co.
6500 Grissom Rd. San Antonio, TX 78238 210 -684-4806

Deep Roots
310 E. Cheryl San Antonio, TX 78228 210 -435-1478

E & E Orchids Nursery
1803 S. E. Loop 410 San Antonio, TX 78220 210-648-5300

E And E Orchids And Greenhouses
4636 Aloha San Antonio, TX 78219 210- 661-4033

Evergreen Garden Center
335 W Olmos #121 San Antonio, TX 78212 210- 738-0225

Fanick's Garden Center Inc.
1025 Holmgreen San Antonio, TX 78220 210-648-1303

Fiesta Nursery
10585 Shaenfield Rd. San Antonio, TX 78254 210-688-3320

Gary's Greenhouse
11317 W. Loop 1604 S San Antonio, TX 78252 210- 622-3344

Green's D Nursery
10076 S. Zarzamora San Antonio, TX 78224 210-628-6050

Greens and Blooms Inc
1550 E Borgfeld Rd San Antonio, TX 78260 210 -494-7004

Hill Country Emporium
12577 Bandera Rd Helotes, TX 78023 210 -695-5240

Landscape Marketplace Inc.
1031 Austin Hwy San Antonio, TX 78209 210-822-1335

Lawns Of Beauty Incorporated
9444 Bandera Road San Antonio, TX 78250 210 -684-2110

Los Patios Inc.
2015 NE Loop 410 San Antonio, TX78217 210 -655-6171

Maldonado Nursery And Landscaping Inc
4393 Stahl San Antonio, TX 78217 210 -599-1219

Milberger Landscaping Incorporated
3920 North Loop 1604 E San Antonio, TX 78247 210 -497-3757

Montgomery Road Nursery
Rt. 9 Box 901 San Antonio, TX 78252 210 -677-8961

More Plants
2014 Fresno San Antonio, TX 78201 210 -271-0858

Oak Island Garden
19120 State Hwy 16 S San Antonio, TX 78264 210 -624-3800

Paul Martinez Nursery
18730 IH 35 Elmendorf, TX 78112 210 -633-2620
Del; Bus. 3 yrs; 33%; Nat; Live Oaks Cedars, Elms; Exp; HoursThur. 1-6, Fri-Sun 9-6, closed M Tu & W

Pauls Desert
200 Schertz Parkway Schertz, TX 78154 210-658-7561

Phil's Nursery- C/O Phil Hovey
11040 S. Zarzamora San Antonio, TX 78224 210 -628-1037

Phil's Nursery
11040 S. Zarzamora San Antonio, TX 78224 210 -628-1037

Plant Interscapes Incorporated
6436 Babcock Rd San Antonio, TX 78269 210 -696-7104

San Antonio Garden Center Incorporated
3310 N. New Braunfels San Antonio, TX 78209 210 -824-9981

Secret Gardens
23966 West IH 10 San Antonio, TX 78252 210-698-4372

Shades Of Green
334 W Sunset Rd San Antonio, TX 78209 210 -822-6344
Del; Bus.20 yrs; 33%; Nat; Color; Exp 2 yrs; Mem; News; Hours M – Sat 9 – 5 Sun 10 – 4

Tapia Tree Sales
10635 W Loop 1604s San Antonio, TX 78252 210-622-3936

The Garden Center Inc
P.O. Box 680983 San Antonio, TX 78268 210 -647-7900

The Grove Nursery
1319 Austin Hwy San Antonio, TX 78209 210 -822-6298

The Secret Garden
555 W Bitters Rd. San Antonio, TX 78216 210-494-4220

Vernons Nursery Incorporated
225 Faith San Antonio, TX 78228 210 -433-9131

Wolfe Nursery
3700 Broadway San Antonio, TX 78209 210-822-7311
7007 San Pedro Ave. San Antonio, TX 78216 210-342-8291
9455 W IH 10 San Antonio, TX 78230 210-641-7559
6214 NW Loop 410 San Antonio, TX 78238 210-681-5813
8802 Perrin Beitel Rd. San Antonio, TX 78217 210-653-9611
6714 S. Flores St. San Antonio, TX 78221 210-922-4321
www.wolfenursery.com

Nursery Specialties Key

Del - Delivery
Bus- Years in operation
% plant stock grown w/in 75 mile radius
Nat - Native plants
Exp - 2 years minimum staff experience
Mem - Member of nursery association
Learn C - Learning center
Fert - Organic fertilizer & pesticide
Demo G - Demonstration garden
News - Newsletter
T Spade - Tree spade

See page 68 for survey details.

903 Area Code - Tyler, Longview, Texarkana, Sherman and Area

A Garden Spot
Rt. 2 Box 850 Hawkins, TX 75765 903-769-9340

Backyard Nature Store
4213 Texoma Pkwy. Sherman, TX 75090 903-893-0599

Baldwin's Patio Yardbirds And Critters
Route 8 Box 1024 Kemp, TX 75143 903-432-4686

Barlow's Nursery
700 West 2nd Ave. Corsicana, TX 75110 903-872-0556

BJ's Perennials
Rt. 1 Box 379B Wills Point, TX 75169 903-873-6269

Blue Angel Gardens
P.O. Box 1473 Mount Vernon, TX 75457 903-588-2903

Blue Moon Gardens
13062 FM 279 Chandler, TX 75758 903-852-3897
Bus. 15yrs; 33%; Herbs & Perennials; Exp; Demo G; News; Fert; Hours M-Sat 9-5:30, Sun 12-5; www.bluemoongardens.com

Bluebonnett Gardens
Route 4, Box 31A Paris, TX 75462 903-785-8175

Bob Wells Nursery
P.O. Box 606 Lindale, TX 75771 903-882-3550

Bratcher's Greenhouse And Nursery
Rt. 1 Box 233 Detroit, TX 75436 903-982-5918

Breedlove Landscape Company
P.O. Box 450 Tyler, TX 75711 903-597-7421

Browns Nursery & Sod
1206 Yorkshire Sherman, TX 75090 903-465-8830

Brown's Plant Farm
P.O. Box 1251 Hawkins, TX 75765 903-769-3567

Buddys Nursery And Garden Center
3407 N. Hwy 259 Kilgore, TX 75662 903-984-7023

Bunch Garden Center
307 Baylor Texarkana, TX 75501 903-838-8525

C & M Landscaping And Nursery
Hwy 98 & FM 1840 New Boston, TX, 75570 903-628-0406

C And R Garden Center
Route 1, P.O. Box 1538 Quitman, TX 75783 903-967-2592

Calverts Garden Center
807 West 16th Street Mt Pleasant, TX 75455 903-572-7414

Canton Garden Center
P.O. Box 878 Canton, TX 75103 903-567-0107

Canton Garden Center Inc
602 S Trade Days Blvd Canton, TX 75103 903-567-0107

Carruth Irrigation And Nursery
P.O. Box 8815 Longview, TX 75607 903-236-7555

Changing Seasons
Rt. 4 Box 305 | Bonham, TX 75418 | 903-640-4211

Chittys Garden Center
817 Houston Street | Wills Point, TX 75169 | 903-873-2553

Coopers Country Greenhouse
5147 Mark's Hill Rd | Marshall, TX 75672 | 903-938-7173

Corsicana Geranium Gardens
P.O. Box 1871 | Corsicana, TX 75151 | 903-874-5989

Country Gardens Herb Farm & Nursery
RFD 1 Van A | Sherman, TX 75090 | 903-482-5170

County Landscapes
4542 CR 2624 | Caddo Mills, TX 75135 | 903-527-4129

Creekside Garden Shop
2111 Hwy 66 | Caddo Mills, TX 75135 | 903-527-4731

Cycle Of Season
400 N. Main St. | Lindale, TX 75771 | 903-882-5666

Daniels Nursery
Rt. 2 Box 15 | Bonham, TX 75418 | 903-583-4741

Davis Plant Farm
P.O. Box 9 Hwy 79 | New Summerfield, TX 75780 | 903-726-3473

Deb' S Plants
Rt. 4 Box 775 B | Jacksonville, TX 75766 | 903-586-1765

Diamond M Nursery
P.O. Box 293 | Winnsboro, TX 75494 | 903-629-3591

Donovan's Landscape And Garden Center
7843 US Hwy 259 S. | Henderson, TX 75654 | 903-898-2891

Dyess Florist And Garden Center
1923 S Jackson | Jacksonville, TX 75766 | 903-586-6214

East, TX Garden Center
2011 Gilmer Rd. | Longview, TX 75604 | 903-297-2778

Emerald Gardens Center Inc
5006 Judson Rd. | Longview, TX 75605 | 903-663-4757

Every Bloomin Thing
P.O. Box 81 | Laneville, TX 75667 | 903-863-5621

Farm And Garden Center
3005 Stone Road | Kilgore, TX 75662 | 903-984-6110

Four Seasons Produce And Nursery
Rt. 1 Box 3994 | Daingerfield, TX 75638 | 903-645-7060

Freeman Nurseries Incorporated
2501 New Boston Road | Texarkana, TX 75501 | 903-793-1392
Bus. 76 yrs. 33%; Nat; Exp; Mem; Learn C; Fert

Garden Valley Farm And Garden Center
Rt. 2 Box 2348 | Quitman, TX 75783 | 903-967-2212

Gardenland Nursery
810 Frisco Rd | Sherman, TX 75090 | 903-868-1938

Green Acre Gardens
P.O. Box 807 | Gilmer, TX 75644 | 903-725-5421

Green Acres Plant Farm Inc
Rt. 9 Box 36 | Jacksonville, TX 75766 | 903-586-5011

Groom's Discount Nursery
12285 CR 2907 | Eustace, TX 75124 | 903-887-8561

Growing Seasons
3206 New Boston Rd | Texarkana, TX 75501 | 903-832-7988

Guinn Plant Farm
Route 1, Box 316-2 | Jacksonville, TX 75766 | 903-586-0651

H And N Landscaping
P.O. Box 7450 | Texarkana, TX 75505 | 903-792-0831

Hardman Nursery
507 Kickapoo | Frankston, TX 75763 | 903-876-4231

Harris Nursery Inc
10940-CR 490 | Tyler, TX 75706 | 903-593-8617

Haws Nursery Inc.
2011 Gilmer Rd. | Longview, TX 75604 | 903-297-2778

Hickory Hill Gardens
P.O. Box 176 | Avinger, TX 75630 | 903-562-1391

Hillside Garden Sales
NC 4 Lake Cherokee | Longview, TX 75603 | 903-643-7051

Hughes Plant Farm Inc
11196 FM 34W | Bullard, TX 75757 | 903-894-7737

Hutto's East 80 Garden Center
4196 E US Hwy 80 | Longview, TX 75601 | 903-757-6394

Hwy 259 Plants
P.O. Box 72 | Ore City, TX 75683 | 903-968-4142

Ingram Nursery
12549 Highway 64 West | Tyler, TX 75704 | 903-592-7034
Del; Bus.45yrs; Azaleas, Camellias; Exp; Mem; Fert;
M-F 8-5 Sat 8-12 (1 in Sprg), Sun closed.

J R Lockhart And Son Nursery
15291 Hwy 155 S | Tyler, TX 75703 | 903-561-0276

Jaden's Garden
P.O. Box 1461 | Gilmer, TX 75644 | 903-734-4306

JD's Plant Farm And Landscaping Inc.
4812 Elysian Fields Rd | Marshall, TX 75670 | 903-835-9989

Jerry Cooper's Plants And Greenhouse Mfg.
252 PR 2126 | Marshall, TX 75672 | 903-938-5020

Josefina Roses
805 Walcott | Jefferson, TX 75657 | 903-665-9271

Lakeside Greenhouse
1133 Murray Lake | Nash, TX 75569 | 903-832-2967

Lamar Gardens Landscape Center
2120 N FM Rd. 1417 | Sherman, TX 75090 | 903-870-7066

Langley's Green Things
125 Amis Rd | Texarkana, TX 75501 | 903-832-7006

Lindsey Nursery
1231 FM 279 | Ben Wheeler, TX 75754 | 903-833-5842

Little Country Greenhouse
Route 2, Box 395-A | New Boston, TX 75570 | 903-628-2991

Lone Star Rose Nursery Inc.
14714 Highway 64 W. | Tyler, TX 75704 | 903-592-8161

Mac's Nursery And Landscaping
Rt. 5 Box 53 | Carthage, TX 75633 | 903-622-4719

Martin Springs Plant Farm
HCO 1 Box 184 | Sulphur Springs, TX 75482 | 903-885-5734

Marys Nursery
450 Live Oak | Canton, TX 75103 | 903-567-2834

Neal Nursery
P.O. Box 1694 | Lindale, TX 75771 | 903-882-6960

Palestine Garden Center
206 Alexander St | Jacksonville, TX 75761 | 903-589-3922

Patrick Farm
13311 FM 849 | Lindale, TX 75771 | 903-882-4844

Pea Shed Plant Co
213 Sheridan Dr | Palestine, TX 75801 | 903-729-6349

Petty's Nursery
Rt. 6 Box 280a | Clarksville, TX 75426 | 903-427-5564

Pickett's Plants And Produce
4387 FM 2795 | Alba, TX 75410 | 903-473-4430

Pineda's Nursery
2001 W 2nd Ave | Corsicana, TX 75110 | 903-872-3793

Pines Garden Center And Nursery
Rt. 1 Box 1106 | Marietta, TX 75566 | 903-835-2901

Plant World
P.O. Box 215 | Fairfield, TX 75840 | 903-389-4822

Premier Landscapes & Nursery
5105 McKnight Rd. | Texarkana, TX 75503 | 903-831-5003

Prices Golden Growers
Hwy 59 S | Texarkana, TX 75501 | 903-832-3939

Nursery Specialties Key

Del - Delivery
Bus- Years in operation
% plant stock grown w/in 75 mile radius
Nat - Native plants
Exp - 2 years minimum staff experience
Mem - Member of nursery association
Learn C - Learning center
Fert - Organic fertilizer & pesticide
Demo G - Demonstration garden
News - Newsletter
T Spade - Tree spade

See page 68 for survey details.

Rhoades Nursery
Rd. 4008 | New Boston, TX 75570 | 903-628-2366

Rose's Nursery And Landscaping
Rt. 2 Box 332 | Yantis, TX 75497 | 903-342-7710

Roses USA
1625 S. Main St. | Tyler, TX 75706 | 903-882-6193

Rozell Rose Nursery
Highway 31 W. | Tyler, TX 75709 | 903-595-5137

S And S Nursery
9982 RP Rd. 3815 | Quinlan, TX 75674 | 903-356-4365

Secret Gardens Florist And Plants
315 Kelly Ln. | Seven Points, TX 75143 | 903-432-1919

Senter Nursery
P.O. Box 305 | Whitehouse, TX 75791 | 903-839-2626

Shade Tree Garden
1328 Lakeway Dr. | Chandler, TX 75758 | 903-849-3992

Shimic's Lawn And Garden Center
Rt. 4 Box 1128 | Jefferson, TX 75657 | 903-755-2295

Silk Tree Farm
1984 Silk Tree Rd | Gilmer, TX 75644 | 903-734-3119

Silver Lake Garden Center
Route 4 Box 103d | Grand Saline, TX 75140 | 903-569-2301

Skidmore's Nursery And Landscaping
Rt. 1 Box 145 | Pattonville, TX 75468 | 903-784-8629

Stevens Garden Center
515 S First St | Sherman, TX 75090 | 903-892-4125

Steves Nursery
4386 Hwy 34 South | Greenville, TX 75402 | 903-883-2911

Stinson's Nursery
FM 2253 | Texarkana, TX 75501 | 903-832-8588

T And R Nursery
12652 Hwy. 155 S | Tyler, TX 75703 | 903-561-3917

Terry's Plant Farm
P.O. Box 333 | Whitehouse, TX 75910 | 903-839-4559

Texoma Lanscapes And Garden Center
5020 Texoma Parkway | Dennison, TX 75020 | 903-892-3908
Bus 17yrs; 33%; Nat; Texas Natives; Exp; Demo G; Learn C; News; Fert; Open by appt only.

The Flower Bed
P.O. Box 871 | Whitewright, TX 75491 | 903-364-9974

The Garden Center
4701 Stonewall | Greenville, TX 75401 | 903-454-6654

The Garden Center
201 A W I30 | Mt Vernon, TX 75457 | 903-588-2454

The Garden Gate
7429 Vz Cr 1507 | Grand Saline, TX 75140 | 903-569-1460

The Garden Shop
P.O. Box 692 Frankston, TX 75763 903-876-5056

The Plant Place
3107 Stone Rd Kilgore, TX 75662 903-984-7931

The Produce Shed
13353 Cr 46 Tyler, TX 75704 903-595-5383

The Seasons
287 Cr 402 Carthage, TX 75633 903-693-8482

The Secret Garden
800 N Greer Pittsburg, TX 75686 903-856-0207

The Secret Garden
9816 FM 751 Quinlan, TX 75474 903-356-5031

Turney Plant Farm Retail #2
P.O. Box 842 Jacksonville, TX 75766 903-586-8675

Twin Oaks Landscaping & Nursery
2107 Highway 691 West Denison, TX 75020 903-4632205

Twitty Nursery
4815 W. 7th St. Texarkana, TX 75501 903-838-5591

Tyler Rose Nursery
13877 Cr. 431 Tyler, TX 75706 903-882-6193

254 Area Code - Waco, Temple Area

Abdon Olvera
728 Nowlin Mexia, TX 76667 254 -562-5439

Annie's Garden And Gifts
513 Kemp St Cameron, TX 76520 254 -697-2170

Barrera Landscaping Supply
P.O. Box 445 Hewitt, TX 76643 254- 666-0060

Billy Bud's Flowers Joyce Woodward
P.O. Box 34 Ross, TX 76684 254- 829-2928

Bonnie's Greenhouse
5198 Orchard Lane Waco , TX 76705 254-799-7909

Bradberry's Best Inc
621 S Patrick Dublin, TX 76446 254- 445-2388

Cen Tex Nursery
P.O. Box 665 Temple, TX 76503 254- 773-5191

Central TX Horticultural Services
2909 Forest Trail Temple, TX 76502 254- 773-0836

Circle R Nursery
Rt. 1 Box 8501 Whitney, TX 76692 254 -694-6274

College Hill Nursery And Landscape
Route 3 Box 299A Hillsboro, TX 76645 254 -582-2637

Country Colors II
804 North Lacy Drive Waco, TX 76705 254 -799-8634

Country Nursery
14101 FM 1105 Jarrell, TX 76537 254 -527-3787

Cove Nursery And Landscaping
407 West Hwy 190 Copperas Cove, TX 76522 254- 547-5888
Del; Bus.17yrs; Nat; Mem; Demo G; Learn C; Fert.

Creekside Nursery
P.O. Box 547 Hamilton, TX 76531 254- 386-5157

Elkins Nursery
Rt. 7 Box 7130 Belton, TX 76513 254 -939-2482

Garden Gate
1733 Ross Rd. Elm Mott, TX 76640 254- 829-2484

Green Thumb Nursery
P.O. Box 13 Teague, TX 75860 254- 739-3658

Greenleaf Nursery And Landscaping
401 W. Ctrl Tx Expwy. Harker Heights, TX 76548 254 -699-4600

Greenlife Nursery Greenlife Nursery
P.O. Box 8355 Waco, TX 76714 254 -776-2400

Greenmaker Nursery
P.O. Box 203 Stephenville, TX 76401 254 -965-7273

Harris Creek Farms
Church Rd. Waco, TX 76706-3279 254-848-2103

Hill Country Cottage Gardens
13261 Blackberry Salado, TX 76571 254 -947-0416

Homestead Gardens
Rt. 1 Box 15B Covington, TX 76636 254 -854-2322

Jeff's Nursery And Landscape
1214-N Ave G Clifton, TX 76634 254 -675-3225

LakeBelton Nursery
474 Centex Sportsman Rd. Belton, TX 76513 254 -780-1353

Lee And Lea's Nursery
702 W Cntrl Tx Expwy. Harker Heights, TX 76548 254 -690-8844

Miller Nursery And Tree Company
Route 1 Box 61 Stephenville, TX 76401 254- 968-2387
Del; Bus.20yrs.; 33%; Nat; Mem; T Spade.
www.miller@ourtown.com

Natural Garden Organics
20760 Hwy 36 Gustine, TX 76455 254-386-3204

Organic +
10568 N River Crossing Waco, TX 76712 254- 848-2103

Pancake Nursery
Route 1 Bartlett, TX 76511 254- 527-3505

Radle's Nursery
Lady Bird Road Waco, TX 76712 254-848-5300

Robinson Drive Greenhouses
628 N. Robinson Dr. Waco, TX 76706 254-662-0311

Storm's Nursery
9605 China Springs Rd., Waco TX 76708 254-836-0022
Del,; Bus.4 yrs., Nat, Trees & Native Plants; Exp 2 yrs; Mem; Learn C; Fert; Hours M-Sat 8 – 6, Sun 10 – 5 clsd. Thanksgiving, New Year, Xmas.

Texas Gardens
400 Park Rd. 4 S Burnet, TX 78611 254-547-7035

The Nursery
6363 FM 439 Belton, TX 76513 254 -389-5836

Wolfe Landscapes
1124 Fairway Road Waco, TX 76712 254-772-7890

940 Area Code - Wichita Falls, Denton, North Central Texas and Area

Agricultural Services
Rt. 1 Box 54 Bowie, TX 76230 940-872-4896

Bowie Nursery
Rt. 5 Box 730 Bowie, TX 76230 940-872-1491

Circle J Farmers Market
3525 Alta Mesa Fort Worth, TX 76133 940-769-3590

Crook Fish & Nursery
211 Avenue F NE Childress, TX 79201 940-937-3610

Four Seasons
3333 E. University Dr. Denton, TX 76208 940-566-2172

Gardens Accents
Rt. 1 Box 281 Graham, TX 76450 940-549-7219

Harris Nursery And Landscape
3209 Lawrence Rd Wichita Falls, TX 76308 940-691-3145

Hilltop Nursery
246 W Hwy 380 Bridgeport, TX 76426 940-683-4682

Holt Nursery And Landscape
3913 Kell Blvd. Wichita Falls, TX 76308 940-691-4757

Jacksboro Garden Center
124 S Main Jacksboro, TX 76458 940-567-6699

Loneoak Nursery And Landscaping
1403 Chico Hwy Bridgeport, TX 76026 940-683-8006

Lovelace Nursery, Landscape & Tree Service
P.O. Box 39 Sanger, TX 76266 940-458-5674

Meador Nursery
2612 Fort Worth Drive Denton, TX 76205 940-382-2638

Nolen's Resale Nursery Floral
601 S Graham In Rear Henrietta, TX 76365 940-538-5828

P and C Country Greenery
P.O. Box 364 Boyd, TX 76023 940-433-5800

Renshaw Greenhouse & Landscaping
415 West Mill Decatur, TX 76234 940-627-3453

Scott Street Lawn And Garden
900 S. Grand Gainesville, TX 76240 940-665-8968

Shade Tree Nursery And Landscaping
3122 Iowa Park Rd Wichita Falls, TX 76306 940-322-9833

Smith Gardentown
4100 Kemp Blvd. Wichita Falls, TX 76308 940-692-7100

Smith's Greenhouses Inc
4100 Kemp Blvd Wichita Falls, TX 76308 940-692-7100

Steve Smith Landscape And Nursery
5001 S. I-35 E Denton, TX 76205 940-382-3169

Texoma Homes Garden
P.O. Box 1414 Vernon, TX 76385 940-552-2900

Tyler's Natural Garden Landscape Company
143 Tower Hill Ln. Gainsville, TX 76240 940-665-6260

979 Area Code

Borden Garden Center
Rt. 3 Box 38 Weimar, TX 78962 979-733-0590

Country Greenhouses
P.O. Box 533 Lexington, TX 78947 979-773-4837

Davida'r Secret Garden
302 Duson Palacios, TX 77465 979-245-2981

936 Area Code - Livingston and Plantersville

Cooks Nursery
Route 14 Nacogdoches, TX 75964 936-564-6359
Bus 43yrs; Azaleas, geraniums, pointsettias, seasonal;
Hours Mon- Sat 9-5:30, Sun. 1-5

Timber Lake Nursery And Stone Inc
Rt. 9 Box 98C Livingston, TX 77351 936-328-5875

All Around Garden Center
240 Renfaire Dr. Plantersville, TX 77363 936-894-2706

903 Area Code- Texarkana

Prices Golden Growers
Hwy 59 S Texarkana, TX 903-832-3939
Home grown plants

Survey Purpose

Landscape Guides LLC conducted two separate surveys of the nurseries and garden centers. The purpose of the surveys was to obtain more information about nurseries and garden centers. The following questions were included in the survey:

Number of years in operation?
Do you provide delivery service, do you stock native plants consisting of more than 25 tree or shrub species?
Do you have a reputation for certain plants?
Do you have an in-house landscape architect or designer?
Do your employees average 2 or more years of experience in the industry?
Do you have a web site?
Do you have a demonstration garden or instructional display?
Do you have a reference or learning center?
Do you have a newsletter?
Do you carry organic pesticides or fertilizers?
Do you have a tree spade for transplanting larger trees?

Sod Farms

Texas

Abilene 915:

Henry's Grass Co.
1945 N. Treadway Abilene, TX 79601 673-8811

Amarillo 806:

N Bar H Turf Farms
6270 S. Osage Amarillo, TX 79118 662-0861

Austin 512:

AAA Grass and Landscaping Inc.
5910Highway 290 West Austin, TX 78701 512-892-3636

Afs Environmental Inc.
12201 Roxie Dr., Austin, TX 78729 512-918-0438

Austin Grass
1809 Ouida Dr. Austin, TX.78728 512-251-5222

Austin Landscape Supplies
16407 fm 1325 (Burnet Rd.) Austin, TX 78728 512-255-2517

Austin Landscape Supplies
5317 S. IH 35 Georgetown, TX 78262 512-930-2311

CedarPark Turf Grass & Garden
1715 W. Whitestone Blvd. Austin, TX 78745 512-250-1885

Daniel Stone and Landscaping Supplies
12000 Hwy 290 W Austin, TX 78737 512- 288-5657

Don's Grass
6408 Brodie Ln. Austin, TX 78745 512-892-5620

Farmer's Grass
1305 Leander Dr. Austin, TX, 78628 1-800-775-7637
512-259-0099

Fuquay Inc.
3000County Rd. 175 Leander, TX 78641 512-260-9899

Quality Grass and Landscaping
5109 Ben White Blvd. Austin, TX 78741 512-447-5200

Southwest Turf and Erosion Products
12140 FM 969 Austin, TX 78724 512-276- 7446

Tischer Zoysia Grass Farm Inc.
5705 Susie Ct. Austin, TX 78757 512-452-1230

254 Area Code - Beaumont

Barrera Landscaping & Supply
201 Sun Valley Blvd. Hewitt, TX 76643 666-9806

Doguet Grass Farm (China Ditch Farms)
795 S. Major Dr. Beaumont TX 77630 409-886-2297

Texas Turf & Trees
4935 Belton Lane Ext. Anahuac, TX 77514 409-267-6000

Corpus Christi 361:

Grass Co. Inc.
7730 Bay Dr. Corpus Christi, TX 78362 361-993-3700

Joseph's Nursery
6051 Ayers St. Corpus Christi, TX 78415 361-851-0401

Fox Tree and Landscaping Nursery
5949 La Costa Corpus Christi, TX 78413 361-939-7017

Dallas & Ft. Worth:

A1 Grass Co.
8463 FM 1565 Royce City, TX 75189 972-423-8814

Anderton , Jimmie Grass Co.
3315 Cr. 2546 Quinlan , TX

Arlington Grass Co.
2417 S. Cooper St. Arlington, TX 76015 817-261-7747

Brigance , Loyd Grass Sales Inc.
11700 Hwy 121 Plano, TX 75024 972-712-3444

Cherokee Grass Co.
I-35 and Danieldale , Dallas, TX 75232 972-228-9007

Dorothy Grass
1022 Stemmons Lewisville, TX 75067 972-219-0607

A-1 Grass Co.
1320 E. 14th St. Plano, TX 75074-6206 972-423-8814

A-1 Grass Co.
8463 FM1565 Royce City, TX 75189

A Systems Designs
360 Old Justin rd. Argyle, TX 76226 940-464-7446

AG Sod Farms Inc.
440 N. Hwy. 377 Aubrey, Tx 76227-9063 940-365-9922

Brigance Loyd Grass Sales, Inc.
11700 Hwy 121 Plano, TX 75024 972-712-3444

Brokers Quality Grass
185 N. Hwy 35 E. Carrolton, TX 75006 972-466-0410

Dorothy's Grass Co.
1022 S. Hwy 35E, 45 Northwood MHP Lewisville TX 75067
972-221-2331

Fort Worth Grass
9409 N. Frwy Fort Worth, TX 76177 817-232-0002

Gardner Turfgrass, Inc.
Hwy 440 N. Hwy 377 Tioga, TX 76271 817-430-8582

Lubbock 806:

Texas Green Turf Farm
99st. and Quaker Lubbock, TX 79416 806-794-2601

San Antonio 210:

Ambassador Enterprises Grass Farm
12455 S. Blue Wing Rd. San Antonio, TX 78223 210-633-2919

Big Green
11031 Wye Dr. San Antonio, TX 78217 1-800-711-3300

Bladerunner Farms
1302 Mauermann Rd. San Antonio, TX 78224 210-628-4164

Coastal Sod Co.
3700 W White Rd. San Antonio, TX 78219 210-648-7005

Del's Landscape Materials Inc.
8587 Bandera Rd. San Antonio, TX 78250 210-681-2744

Douglas W. King Seed Co.
4627 Emil Rd San Antonio, TX 78219 210-661-4191

Earth Materials and Nursery
10660 S E Loop 410 San Antonio, TX 78214 210-928-2583

Fairway Nursery
15135 Herring Road Atascosa, TX 78002 210-622-6262

Fertile Garden Supply
8349 Leslie Rd San Antonio, TX 78254 210- 688-9435

Garden-Ville Landscape Supply
7561 E. Evans Rd San Antonio, TX 78266 210-651-6115
26807 Hwy 281 North San Antonio, TX 210-497-1777
10997 S. Hausman Rd. San Antonio, TX 210-682-7645

Grass Farms
15934 Nacogdoches San Antonio, TX 78247 210- 646-8185

Guajardo Grass and Topsoil Sales
16322 Boulder Pass San Antonio, TX 78247 210-599-0147

Lytle Grass Farm
14744 Jarralt Rd. Lytle, TX 78052 830-709-3338

McTurf Inc.
13411 Vista del Mar San Antonio, TX 78216 493-9480

Milberger Landscaping and Nurseries
3920 N Loop 1604 E San Antonio, TX 78247 497-3760

Sierra Sod Co.
10522 Lake Braunig San Antonio, TX 78223 633-0131

Thomas Brothers Grass
Pottet, TX 800-697-4732

El Paso 915:

Gardner Turfgrass Inc.
5590 McNutt Rd El Paso, TX 79932 589-0401

Royalty Landscaping
3012 Bear Lake Pl El Paso, TX 79936 857-3150

Tyler:

Cannon Grass & Tree Farm
1143 CR 163 & 164 Kaufman, TX 75142 972-962-4697

Miller Grass Co.
10025 Plano Rd. Dallas, TX 75238-1788 214-348-0810

Oakwood Grass Farm
3230 John Carpenter Freeway, Irving, TX 75062 972-554-1104

Taylor Grass and Stone Co.
1120 S. Texas Lewisville, TX 75057 972-436-7973

Tex Turf Sod Co.
7701 Davis Arlington, TX 76013 817-498-2322

Del Rio 361:

Bowers Landscaping and Carpet Grass Farm
101 Clouse Drive Del Rio, TX 78840 361-775-5692

Houston 218:

A Grass Farms Inc.
25615 SW Rosenberg, TX 77471 281-342-0123

AAA Grass Sales (AAA Ventures Inc.)
4911 Spring Cypress Spring, TX 77379 261-376-8440

ABC Sports Turf
2000 S. Dairy Ashford St. Houston, TX 77077 281-493-3327

Allied Grass Farms Inc.
27706 Jeans Spring, TX 77373 281-353-2900

Arnolds Green Acres Grassfarm & Turfarm
1011 Gum Gully Rd. Houston, TX 77088 281-353-2900

Beltway Turf Inc.
11927 N. Garden Houston, TX 77071 713-272-6033

Brown Brothers Grass Nursery
5710 Reed Road Houston, TX 77033 713-733-3344

Champion Grass
1463 Hwy 6 Houston, TX 77077 281-558-9948

Champion Hydro Lawn Inc.
13226 Kaltenbrun Houston, TX 77086 281-445-2614

Champion Top Soil, Inc.
1463 Highway 6 South Houston, TX 77077 281-558-9948

Cornett Grass
9035 FMkt 1960 West Houston, TX 77070 291-890-0733

Crosby Grass Farm		
1231 Krenek	Crosby, TX 77532-6254	281-328-6388
Garden Accents		
14907 Treichel	Tomball, TX 77375	281-351-4804
Grimes Grass Company		
1106 Highway 6 South	Houston, TX 77077	281-493-5505
Fountain and Statuary Etc.		
11804 Hempstead	Houston, 77092	713-957-3672
Houston Grass Co.		
24418 N Ih 45	Spring, TX 77386-1939	281-364-9750
Mata Turf		
10408 Tanner Rd.	Houston, TX 77041	713-896-9532
Milberger Turfgrass		
7763 I-45 Conroe	Houston, TX 77385	281-443-0915 1-800-445-2602
Murf Turf Farm Inc.		
15204 Bohemian	Crosby, TX 77532-6001	281-328-2812
Saddlebrook Inc.		
15930 Herongate	Houston, TX 77084	281-345-1200
Shuman's Grass Farm Inc.		
3901 Sherri Lane	Crosby, TX 77532	281-328-2891
Spring Grass Inc.		
7715 Lquetta Rd.	Houston, TX 77070	281-320-8800
Thomas Bros Grass LTD		
1926 Acton Hwy	Grandbury, TX.76049	972-219-5900

Waco 254:

Quality Grass		
6913 W. Woodway Drive	Woodway, TX 76712	772-7633
South Bosque Turf Grass and Garden Center		
26005 W. Hiway 84	Waco, TX 76712-4011	848-2411
Waco Turf Grass Producers		
FM 434 7267 S. 3rd St. Rd.	Waco, TX 76706	662-3344

Westlaco 956:

Guajardo Turf farms		
1210 W Expressway 83	Westlaco, Tx 78596	956-968-6313

Wichita Falls 940:

Chickasha Sod and Grass Farm		
Rt 1 Box 211	Amber, OK 73004	940-459-6232

Arkansas

Fayetteville:

Capstick Sod & Lawn Co.		
635 Hiway 71 S.	Fayettville, AR 72701	839-3468
J & R Sod Service		
2402 A Turner St. # A	Springdale, AR 72764	750-3885
Northwest Arkansas Sod & Turf		
174 Killdeer Dr.	Farmington, AR 72730-8619	267-4450
Steve's Sod Store		
125 S. Majestic Rd.	Tontitown, Ark	361-2000
Turfscapes		
218 Apple Blossom Av.	Lowell, AR 72745-9760	770-2111

Fort Smith 501:

Faucher Farms		
5300 McClure Rd	Van Buren, AR 72956	474-7535
Green Tree Landscaping & Tree Service		
8001 S 24th St.	Fort Smith, AR 72908	646-7811
Razorback Sod and Turf		
1408 Mayo Dr.	Burling, AR 72923	452-4788

Little Rock :

Arkansas Garden Center		
15603 Cantrell Rd.	Little Rock, AR 72223	868-9933
Arkansas Sod & Turf Farm		
6200 Crystal Hill Rd N	Little Rock, AR 72118	753-5377
Double A Sod & Turf		
205 Manson Rd.	Sherwood, AR 72120	834-2989
Home Turf		
9109 Moore Ford Rd.	Benton, AR 72015-7301	776-1166
Pleasant Valley Sod Farms		
P.O. Box 157	Houston, AR 72070	759-3088
Quail Valley Farm Inc.		
7509 Cantrell Rd. # 103	Little Rock, AR 72207	975-6281
Winrock Grass Farm Inc.		
905 La Harpe Blvd.	Little Rock, AR 72201	375-0749

Texarkana 870:

Fulton Grass Co.		
791 CR 1	Fulton, AR 71838	870-896-2407

Louisana

Baton Rouge 225:

Couhig Southern Environmental Services		
16312 Perkins Rd.	Baton Rogue, LA 70810	225-752-7378
Don Hart's Landscape Design Inc.		
9163 S. Tiger Bend Road	Baton Rogue, LA 70817	225-753-3865
Mid-South Sod Inc		
69497 Turf Farm Road 1800-852-4244	Abita Springs, LA 70420	505-892-1892
Mr. Sod		
11929 Cloverland Court	Baton Rogue, LA 70809	225-751-2666
Pine Island Turf Farm		
P.O. Box 20202	Beaumont, TX 77720	1-800-683-0363

Southern Sod.		
12640 Hgwy 104	Fairhope, MS 36532	334-928-7277
Woerner Turf		
11770 Airline Hwy	Baton Rouge, LA 71111	504-291-0107

Lafayette 318:

Gator Turf		
217 Des Jardin Ave.	Lafayette, LA 70507	318-896-0076

Shreveport 318:

Ark-La-Tex Farms		
512 Jenkins Road	Shreveport, LA 71107	424-1008
Wetherington Grass Sales		
5301 E. TX	Bossier City, LA 71111	318-747-5274
Wholesale Grassland		
3400 Greenwood Road	Shreveport, LA 71109	635-6820

New Orleans:

Jenkins Soil and Landscape Material		
101 West 33rd. Ave.	Covington, LA 70434	504-892-9118
L & S. Sod Co.		
62 Rosemary Ln.	Picayune, MS 39466	601-798-8819
Perry's Sod		
81430 Bob Baxter Rd.	Bush, LA 70431	504-809-9955

Oklahoma

Durant/Sherman Dennison, Tx 580:

McClure Sod Farms		
Rt 2 Box 245	Hendrix, OK 74741	580-838-2401

Oklahoma City 405:

Allen Farm, Inc.		
476 Allen Road	Norman, OK 73072	405-364-3998
Bentley Turf		
2211 Westpark Drive	Norman, OK 73069	405-364-7511
Canadian Valley Turf Farm		
P.O. Box 424	Spencer, OK 73084	405-771-5135
Sod by Sherry		
17000 W. Foreman Road	Yukon, OK 73099	405-354-7926
Sunbelt Sod Farms		
RT 2 Box 265	Shawnee, OK 748404	405-275-5703

Tulsa 918:

Adams Turf Farm		
13927 E. 171st St	Bixby, OK 74008	918-366-6800
Easton Sod Farm		
P.O. Box 1041	Bixby, OK 74008	918-366-2500
Hilltop Sod Farm		
735 Trower St.	Coweta, OK 74429	918-486-4482
Shirley's Sod Farm		
HC 67, Box 40	Indianola, OK 74442	918-823-4638
Underwood's Black Bottom Sod		
26655 S. 4130 Road	Claremore, OK 74017	918-266-5578
Verdigris Valley Sod Farms		
P.O. Box 97	Claremore, OK 74018	918-341-8742
Wright Sod Farm		
10525 S. 193rd. E Ave.	Broken Arrow,OK 74014	918-455-8373

Botanical Gardens, Arboretums, & Nature Areas

San Antonio Botanical Garden

Dallas Arboritum

Arkansas

Arkansas Territorial Restoration
200 E. 3rd. St. Little Rock, AR 72201 501-324-9351

Camden Driving Garden Tours
314 Adams SW Camden, AR 71701 870-836-6426

Eureka Springs Gardens
US 62 W Route 6 Eureka Springs, AR 72632 501-253-9244

Garvan Woodland Gardens
Univ of Arkansas Hot Springs Nat'l Park, AR 71913 888-530-6873

Jacksonport State Park
P.O. Box 8 Jacksonport, AR. 72075 870-523-2143

Japanese Garden at Civic Center Complex
500 8th St., 8th & State Pine Bluff, AR

Little Rock Zoological Garden
1 Jonesboro Dr. Little Rock, AR 71953 501-666-2406

Mountain Valley Spring Water Co.
150 Central Ave. Hot Springs Nat'l Park, AR 71901 800-643-1501

Peel House Museum & Historical garden
400 S. Walton, Blvd. Bentonville, AR 72712 501-273-9664

Pinnacle Mountain State park
11901 Pinnacle Valley Rd. Roland, AR 72135 501-868-5806

Queen Wilhelmina State Park
3877 Hwy 88 W. HC-07 Mena, AR 71953 501-394-2863

Quality Gladiolus Garden
Arkansas 1 B S, Box 458 Jonesboro, AR 72403 870-932-4533

South Arkansas Arboretum
Mount Holly Road & Timberlane
P.O. Box 7010 El Dorado, AR. 71731 870-862-8131

Louisiana

Acadian Park Nature Trail
E. Alexander St. Lafayette, LA

Acadian Village & Gardens
200 Green Leaf Dr. Lafayette, LA 70506 318-981-2364

Academy of the Sacred Heart
1821 Academy Rd. Grand Coteau, LA 70541 318-662-5275

Audubon State Commemorative Area
Hwy 965
P.O. Box 546 St. Francisville, LA 70775 504-635-3739

Barataria Preserve
7400 Hwy 45 Marrero, LA 70072 504-589-2330

Barnwell Garden & Arts Center
501 Clyde Fant Prkwy. Shreveport, LA 71101 318-673-7709

Beau Fort Plantation
4055 Hwy 494 Natchitoches, LA 71457 318-352-9580

Beauregard Keyes House & Garden
1113 Chartres St. New Orleans, LA 70116 504-523-7257

Briarwood Home of Caroline Dorman
216 Caroline Dorman Rd. Saline, LA 71070 318-576-3379

Burden Research Plantation, Louisiana State University
4560 Essen Lane Baton Rouge, LA 70809 504-763-3990

Catalpa Plantation
St. Francisville, LA 70775 504-635-3372

Cohn Arboretum
12206 Foster Rd. Baton Rogue, LA 70811 504-775-1006

Elsong Garden & Conservatory
2006 Riverside Dr. Monroe, LA 71201 318-387-5281

Ford Municipal Park
S. Shore of Cross Lake | Shreveport, LA

Gallier House
1132 Royal St. | New Orleans, LA 70116 | 504-523-6722

Hilltop Arboretum at Louisiana State University
11855 Highland Road | Baton Rogue, LA 70810 | 504-767-6916

Hodges Gardens
Hwy 171, P.O. Box 900 | Many, LA 71449 | 800-354-3523

Houmas House Plantation & Gardens
40136 Hwy 942 Burnside | Darrow, LA 70725 | 504-522-2266

Independence Park Botanic Garden
7950 Independence Blvd. | Baton Rogue, LA | 225-928-2270

Ira S. Nelson Horticulture Center
University of SW Louisiana
200 E. University Ave. | Lafayette, LA 70503 | 318-231-6000

Jungle Gardens - Avery Island
P.O. Box 126 | New Iberia, LA 70560 | 318-369-6246

Kent Plantation House
3601 Bayou Rapides Rd. | Alexandria, LA 71303 | 318-487-5998

Live Oak Gardens
5505 Rip Van Winkle Rd. | New Iberia, LA 70560 | 318-365-3332

Longfellow-Evangeline State Commerative Area
Route 31, 1200 N. Main St. | St. Martinville, LA 70582 | 318-394-2233

Longue Vue House & Garden
7 Bamboo Rd. | New Orleans, LA 70124 | 504-488-5488

Louisiana Nature Center
Read Blvd. at Joe Brown Park
P.O. Box 870610 | New Orleans, LA 70187 | 504-243-3395

Louisiana Purchase Gardens & Zoo
Tichelli Rd., P.O. Box 123 | Monroe, LA 71210 | 318-329-2400

Louisiana State Arboretum
4213 Chicot Park Rd. | Ville Platte, LA 70586 | 318-363-6289

Louisiana Tech University Arboretum
Reese Hall | Ruston, LA 71272 | 318-257-0211

Mount Hope Plantation
8151 Highland Rd. | Baton Rogue, LA 70808 | 504-766-8600

New Orleans Botanical Garden
1 Palm Dr. | New Orleans, LA 70124 | 504-483-9386

Oak Alley Plantation
Great River Rd., 3645 Hwy 18 | Vacherie, LA 70090 | 504-265-2151

Oak Valley Plantation
Great River Rd., 3645 Hwy 18 | Vacherie, LA 70090 | 504-265-2151

Rosedown Plantation & Historical Gardens
12501 Hwy 10 | St. Francisville, LA 70775 | 504-635-3332

Rose Gardens
City Park Esplanade | New Orleans, LA

San Francisco Plantation
Hwy 44 Great River Rd.
P.O. Drawer AX | Reserve, LA 70084 | 504-535-2341

Tezcuco Plantation
3138 Hwy 44 | Darrow, LA 70725 | 504-562-3929

Ursuline Convent Garden
1100 Chartres St. | New Orleans, LA 70116 | 504-529-3040

Windrush Gardens Louisiana State University
4600 Essen Lane | Baton Rogue, LA 70809 | 504-765-2437

Zemurray Gardens
Hwy 40/23115 Zemurray Gardens | Loranger, LA 70446 | 504-878-2284

The Bivin Garden
¼ mi. W. on SH 11 | Shidler, OK 74652 | 918-793-4011

Cann Memorial Garden Center
1500 E. Grand | Ponca City, OK 74604 | 580-767-0427

Cedarville Botanic Gardens
Davis, OK | 580-369-3224

Forest Heritage Center- Beavers Bend Resort Park
P.O. Box 157 | Broken Bow, OK 74728 | 580-494-6497

George M. Sutton Wilderness Park
920 12th Ave. NE | Norman, OK | 405-366-5472

Hambrick Botanical Garden at The National Cowboy Hall of Fame
1700 NE 63rd St. | OK City, OK 73111 | 405-478-2250

Honor Heights Park
Honor Heights Dr. | Muskogee, OK 74401 | 918-684-6302

Horn Canna Farm
8 mi. N. on Hwy 58; 1 ¼ m. E.
Rt. 1, Box 94 | Carnegie, OK 73015 | 580-637-2327

Jo Allyn Lowe Park
Price Rd. & Locust Rd | Bartlesville, OK 74005 | 918-337-5267

Kerr Arboretum & Botanical Area
Talimena Scenic Dr.; 1 ½ mi. E. of Hwy 259
P.O. Box 577 | Talihina, OK 74571 | 580-567-2326

Lendonwood Botanical Gardens
1308 W. 13th St. | Grove, OK 74344 | 918-786-2938

Martin Park Nature Center
5000 W. Memorial Rd. | OK City, OK 73142 | 405-755-0676

Myriad Botanical Gardens
Corner of Reno & Robinson
1 Myriad Gardens | Oklahoma City, OK 73102 | 405-297-3935

North OK Botanical Garden & Arboretum
1220 E. Grand Ave. | Tonhawa, OK 74653 | 580-628-2220

OK Botanical Garden & Arboretum
S of Virginia Ave.; E of Sangre Rd. | Stillwater, OK | 405-744-5404

OK City Zoological Park
2102 NE 50th St. | OK City, OK 73102 | 405-424-3344

Oklahoma

The Bivin Garden
¼ mi. W. on SH 11 | Shidler, OK 74652 | 918-793-4011

Cann Memorial Garden Center
1500 E. Grand | Ponca City, OK 74604 | 580-767-0427

Cedarville Botanic Gardens
Davis, OK | 580-369-3224

Forest Heritage Center- Beavers Bend Resort Park
P.O. Box 157 | Broken Bow, OK 74728 | 580-494-6497

George M. Sutton Wilderness Park
920 12th Ave. NE | Norman, OK | 405-366-5472

Hambrick Botanical Garden at The National Cowboy Hall of Fame
1700 NE 63rd St. | OK City, OK 73111 | 405-478-2250

Honor Heights Park
Honor Heights Dr. | Muskogee, OK 74401 | 918-684-6302

Horn Canna Farm
8 mi. N. on Hwy 58; 1 ¼ m. E.
Rt. 1, Box 94 | Carnegie, OK 73015 | 580-637-2327

Jo Allyn Lowe Park
Price Rd. & Locust Rd | Bartlesville, OK 74005 | 918-337-5267

Kerr Arboretum & Botanical Area
Talimena Scenic Dr.; 1 ½ mi. E. of Hwy 259
P.O. Box 577 | Talihina, OK 74571 | 580-567-2326

Lendonwood Botanical Gardens
1308 W. 13th St. | Grove, OK 74344 | 918-786-2938

Martin Park Nature Center
5000 W. Memorial Rd. | OK City, OK 73142 | 405-755-0676

Myriad Botanical Gardens
Corner of Reno & Robinson
1 Myriad Gardens | Oklahoma City, OK 73102 | 405-297-3935

North OK Botanical Garden & Arboretum
1220 E. Grand Ave. | Tonhawa, OK 74653 | 580-628-2220

OK Botanical Garden & Arboretum
S of Virginia Ave.; E of Sangre Rd. | Stillwater, OK | 405-744-5404

OK City Zoological Park
2102 NE 50th St. | OK City, OK 73102 | 405-424-3344

OK Heritage Center Gardens
201 NW 14th St. | OK City, OK 73103 | 405-235-4458

Omniplex
2100 NE 52nd St. | OK City, OK 73111 | 405-424-5564

Oxley Nature Center Mohawk Park
5701 E. 36th St. N. | Tulsa, OK 74115 | 918-669-6644

Philbrook Museum of Art
2727 S. Rockford Rd. | Tulsa, OK 74114 | 918-749-7941

Red Bud Valley Nature Preserve
161 E. Ave.; 3.8 mi. N. of IH-44
5701 E. 36th St. N. | Tulsa, OK 74115 | 918-669-6644

Sam Noble OK Museum of Natural History
2401 Chautauqua St. | Norman, OK 73072 | 405-325-4712
www.snomnh.ou.edu

Tulsa Garden Center
2435 S. Peoria | Tulsa, OK 74114 | 918-746-5125

Tulsa Zoo & Living Museum
5701 E. 36th N. | Tulsa, OK 74115 | 918-669-6600

Will Rogers Park
3400 N. W. 36th St. | OK City, OK 73112 | 405-943-2739

Woodward Park
21st. S. & Peoria | Tulsa, OK | 918-596-7877

Woolaroc Museum, Wildlife Preserve & Garden
12 mi. SW on Hwy 123, Rt. 3 | Bartlesville, OK 74003 | 918-336-0307

Texas

TX Panhandle:

Amarillo Botanical Gardens
1400 Streit Drive | Amarillo, TX 79106 | 806-352-6513

Don Harrington Discovery Center Botanical Garden
1200 Streit Dr. | Amarillo, TX 79106 | 806-355-9547

North Central Texas:

Dallas Arboretum and Botanical Garden
8625 Garland Road | Dallas, TX 75218 | 214-327-8263

Dallas Garden Center
State Fair Park | Dallas, TX 75210 | 214-428-7476

Dallas Nature Center
7171 Mountain Creek Parkway | Dallas, TX 75249 | 972-296-1955

Dallas Zoo
650 South RL Thornton Frwy. | Dallas, TX 75203 | 214-670-5656

Chandor Gardens
710 W. Simmons | Weatherford, TX 76086 | 817-613-1700

Hermosa Valley Gardens
9410 Hermosa Drive | Dallas, TX 75218 | 214-320-3020

Fort Worth Botanical Garden
Trinity Park
3220 Botanic Garden Dr. | Fort Worth, TX 76107 | 817-871-7686

Fort Worth Water Gardens
1501 Commerce St. | Fort Worth, TX 76102 | 817-871-7698

Garden of Peace
Forest Park Cemetery | Greenville, TX 75403 | 903-455-1510

Mathews Prairie Preserve
W on US 380; N on FM 903 | Greenville, TX

Samuel Grand Municipal Gardens
6200 E. Grand Ave. Dallas, TX 75223

Tankerley Gardens
Tankersley Road Mount Pleasant, TX 75477 903-572-0567

The Heard Native Plant Garden
1 Nature Place McKinney, TX 75069 972-562-5566

North East Texas:

Cargill Long Park
Hollybrook St. Longview, TX 75606 903-753-3281

East TX Arboretum & Botanical Society
Off US 175 on Patterson Road Athens, TX

Helen Lee Estate Daffodil Gardens
21600 County Road 3103 Gladewater, TX 75647 903-845-5780

SFA Mast Arboretum
Wilson Drive Nacogdoches, TX 75962 936-468-4343

Tyler Rose Gardens
420 S. Rose Prkwy Tyler, TX 903-882-6193

Central Texas:

Austin Nature & Science Center
301 Nature Center Dr. Austin, TX 77846 512-327-8181
Antique Rose Emporium
10,000 Hwy 50 Brenham, TX 77833 979-836-5548

Cameron Park Zoo
1701 N 4th St. Waco, TX 76707 254-750-8400

Clear Springs Aviaries & Zoological Gardens
5686 IH-35 S. New Braunfels, TX 78312 830-606-6029

Earl-Harrison House & Gardens
1901 N. 5th St. Waco, TX 76708 254-753-2032

Fredricksburg Herb Farm
402 Whitney Street Fredricksburg, TX 78624 830-997-8615

Lindheimer Home
491 Comal St. New Braunfels, TX 78130 830-625-8766

Lucile Halsell Conservatory
555 Funster San Antonio, TX 78209 210-821-5115

Mayfield Park & Preserve
3808 W. 35th St.
200 S. Lamar Boulevard Austin, TX 78704 512-499-6700

The Lady Bird Johnson Wildflower Center
4801 La Cross Avenue Austin, TX 78725 512-292-4200

Riverside Nature Center
150 Francisco Lemos St.
P.O. Box 293970 Kerrville, TX 780029 830-257-4837

San Antonio Botanical Gardens
555 Funston PL San Antonio, TX 78209 210-207-3263

San Antonio Zoological Gardens
3903 N. Saint Mary's St. San Antonio, TX 78212 210-734-7184

Sutherland's Cactus Garden
N FM 907 & FM 495 Alamo, TX 78516 956-787-2040

TAMU Horticulture Gardens, TX A&M University
Hensel Drive/Dept of Horticulture Sciences, MS 2133
College Station, TX 77843 979-845-3658

Umlauf Sculpture Garden
605 Robert E. Lee Road Austin, TX 78704 512-445-5582

Walnut Springs Memorial Rose Garden
Donegan & Travis Streets Seguin, TX

Wildseed Farms
7 mi. E. on US 290 Fredricksburg, TX 78624 800-848-0078

Westcave Preserve
½ mile past bridge over Pedernales R. on RR 32
Cypress Mill Marble Falls, TX 78654 830-825-3442

Zilker Botanical Garden
2220 Barton Springs Road Austin, TX 78746 512-477-8672

Southeast Texas:

Bayou Bend Collection & Gardens
1 Westcott Street Houston, TX 77007 713-629-7750

Beaumont Botanical Gardens & Warren Loose Conservatory
Fannett at Tyrrell Park Rd Beaumont, TX 77726 409-842-3135

Big Thicket National Preserve
7 miles north on FM 420
3785 Milam Beaumont, TX 77701 409-246-2337

Cockrell Butterfly Garden
1 Hermann Circle Drive Houston, TX 77030 713-639-4600

Cullen Sculpture Garden
1001 Bissonnet Houston, TX 77005 713-845-1000

Hilltop Herb Farm
16 mi. E. on FMM 787
1 Country Ln. Cleveland, TX 77327 281-592-5859

Houston Arboretum & Nature Center
4501 Woodway Drive Houston, TX 77024 713-681-8433

Houston Garden Center & Japanese Garden
Hermann Park- 1500 Hermann Dr. Houston, TX 77004

Houston Zoological Gardens
1513 N. McGregor Houston, TX 77030 713-284-1354

Lilypons Water Gardens
839 FM 1489 Brookshire, TX 77423 281-391-0076

Mercer Arboretum & Botanical Gardens
22306 Aldine Westfield Rd. Humble, TX 77338 281-443-8731

Moody Gardens
1 Hope Blvd. Galveston, TX 77554 800-582-4673

Shimek's Gardens
County Road 237/Rt. 1 Box 267 Alvin, TX 77511 281-331-4395

South Texas:

Bird Garden & Wetlands Pond
Rest area E. of IH-35 N   Rockport, TX

Corpus Christi Botanical Gardens
8545 S. Staple Street   Corpus Christi, TX 78413   361-852-2100

Glady's Porter Zoo
500 Ringgold Street   Brownsville, TX 78520   956-546-2177

Lower Rio Grande Valley Nature Center
301 S. Border in Gibson Park
P.O. Box 8125   Weslaco, TX 78596   956-682-1517

McAllen Botanical Gardens
W. McAllen on Bus. US 83
P.O. Box 220   McAllen, TX 78502   956-562-5566

West Texas:

Big Bend National Park
P.O. Box 196   Big Bend National Park, TX 79834   915-477-2236

Chihuahuan Desert Gardens
Corner of University & Wiggins
Centennial Museum, University of TX at El Paso   El Paso, TX 79968
915-747-5565

Chihuahuan Desert Research Institute
TX 118, 4 miles S/Box 905   Fort Davis, TX 79734   915-364-2499

Davis Mountains State Park
TX 118, Park Road 3
P.O. Box 1458   Fort Davis, TX 79734   915-426-3337

Lubbock Memorial Arboretum
Clapp Park
4111 University Ave.   Lubbock, TX 79413   806-797-4520

Maxey Park Zoo
IH-20 Ex. 40 Access Rd.   Pecos, TX

Mc Murry Garden
McMurry University   Abilene, TX 79605   915-691-6200

Rio Concho Garden
Rio Concho Park   San Angelo, TX   800-375-1206

Sunken Garden Park
W. Avenue D & S Abe St.   San Angelo, TX   800-375-1206

Azeleas in full bloom at Honor Heights Park - Muskogee, OK

Seasonal Calendar

Some years our climate is fairly predictable with temperatures and rainfall close to the average, but don't count on it! The seasoned gardener of the south-central States knows that climate is more unpredictable than not. Be flexible in your planning and planting. Remember the saying...'make hay while the sun shines'.

Spring

- Plant selection is best in the spring! Stock up while the selection is good.
- Plant containerized or b & b trees and shrubs.
- Plant flowers, groundcover, and summer flowering bulbs.
- Apply compost and organic fertilizers around plants.
- Apply soil amendments as recommended from soil test results.
- Spread compost and non-burning fertilizers around plants.
- Lay sod or plant grass plugs in March for zones 8 & 9 and in April for zones 6 & 7.
- Fertilize warm season grasses (St. Augustine, Bermuda, Zoysia, Buffalo, and Centipede.
- Aerate lawn.

Blue Hyacinths

MARCH

SUN	MON	TUE	WED	THU	FRI	SAT

EXPENSES

DATE-	DESCRIPTION-	AMOUNT-

APRIL

SUN	MON	TUE	WED	THU	FRI	SAT

EXPENSES

DATE-	DESCRIPTION-	AMOUNT-

MAY

SUN	MON	TUE	WED	THU	FRI	SAT

EXPENSES

DATE-	DESCRIPTION-	AMOUNT-

Summer

JUNE						
SUN	MON	TUE	WED	THU	FRI	SAT

EXPENSES

DATE-	DESCRIPTION-	AMOUNT-

- June - Fertilize grasses and raise mower blade at a higher setting for cutting shade grasses.
- Keep Hummingbird feeders full of nectar.
- Remove faded flowers (deadhead) and cut back rangy growth to encourage more blooms.
- Apply compost around plants.
- Mulch around new plants to help moderate soil temperatures and conserve moisture.
- Water as necessary (see WATER section)

JULY						
SUN	MON	TUE	WED	THU	FRI	SAT

EXPENSES

DATE-	DESCRIPTION-	AMOUNT-

Ajuga

AUGUST						
SUN	MON	TUE	WED	THU	FRI	SAT

EXPENSES

DATE-	DESCRIPTION-	AMOUNT-

Autumn

Caddo Maple

- Plant shade trees. Root growth and development takes place while there is no new growth above ground.
- Sow shady lawn grass seed mix. This allows for substantially more roots to grow than if planted in the spring.
- Aerate grass as this will allow roots to grow deeper and extend out more.
- Divide perennials like iris, daisies, and daylilies. Dig up bulbs such as gladiolas or dahlias.
- Sow perennial seed or plant new plants.
- Plant spring flowering bulbs such as tulips, daffodils, hyacinth, and crocuses.
- Apply compost and organic fertilizer as necessary.
- Apply soil amendments like sulfur or lime to lower or raise pH as recommended by soil test results.
- November - mulch up to the crown on plants that need protection from the freezes and thaws of winter such as hybrid tea roses.
- November - fertilize trees with spikes or dig cores around trees and apply fertilizer.

SEPTEMBER

SUN	MON	TUE	WED	THU	FRI	SAT

EXPENSES

DATE-	DESCRIPTION-	AMOUNT-

OCTOBER

SUN	MON	TUE	WED	THU	FRI	SAT

EXPENSES

DATE-	DESCRIPTION-	AMOUNT-

NOVEMBER

SUN	MON	TUE	WED	THU	FRI	SAT

EXPENSES

DATE-	DESCRIPTION-	AMOUNT-

Winter

- Bare-root (packaged and lower cost) roses, fruit trees, and shrubs are available in stores in January and early February. Store in a cool low light non-freezing location. When the ground is workable soak the roots in a bucket of water overnight and plant early the next day.

- Keep wild bird feeders full. This is the most difficult time of the year for birds and other wildlife to find food.

- Sharpen hand tools and have power tools serviced before there is a spring backlog.

- Be sure and water plants if the soil moisture levels are low.

- Test the soil in your yard to see if you need amendments.

- Plant seeds indoors for later planting after all danger of frost is past.

- Prune fruit trees and spray with a dormant oil before the buds swell.

American Beautyberry

DECEMBER

SUN	MON	TUE	WED	THU	FRI	SAT

EXPENSES

DATE-	DESCRIPTION-	AMOUNT-

JANUARY

SUN	MON	TUE	WED	THU	FRI	SAT

EXPENSES

DATE-	DESCRIPTION-	AMOUNT-

FEBRUARY

SUN	MON	TUE	WED	THU	FRI	SAT

EXPENSES

DATE-	DESCRIPTION-	AMOUNT-

Your Site Plan

The **Site Plan** provides the basis for understanding what, why, and where. There are two main analysis used in compiling the information for a site plan. 1) Site analysis – is a survey and inventory of all the existing features that might affect the landscape design. This includes factors such as north, south, east, and west locations, prevailing winds, drainage, slopes, and existing structures or objects (utilities, trees, walks, etc.). 2) Use analysis – what is the overall theme you want to achieve, what are traffic patterns, future additions or projects, and do you want defined areas or 'garden rooms'. *Use the grid below to plot your landscape plan.*

Suggested Reading List

Suggested Reading List:

Brown, Robert & Gillespie, Terry. 1995. Microclimate Landscape Design. New York: John Wiley & Sons, Inc.

Chaplin, Lois Trig. 1994. The Southern Gardener's Book of Lists. Dallas, Texas: Taylor Publishing Co.

Clote Jr., Tom. 1987. A Guidebook to Oklahoma Landscaping. Norman, Ok.

Dansereau, Pierre. 1973. Inscape and Landscape. The Human Perception. New York: Columbia University Press.

Duffield, Mary, and Jones, Warren. 1981. Plants for Dry Climates How To Select, Grow and Enjoy. Tuscon, Ariz: H. P. Books.

Garret, Howard. 1993. Texas Organic Gardening Book. Houston, Texas: Gulf Publishing Co.

Heriteau, Jacqueline. 1996. Glorious Gardens. New York: Roundtable Press.

Klingaman, Gerald. Hardiness of Selected Arkansas Landscape Plants. Cooperative Extension Service, University of Arkansas. FSA2097-6M-6-99R.

Helmer, Jane, and Hodge, Karla. 1994. Pictorial Guide To Perennials. Kalamazoo, Michigan: Merchants Publishing Co.

Joyce, David. 1998. The Perfect Plant. New York: Stewart, Tabori, and Chang.

Little, Elbert. Forest Trees Of Oklahoma. Oklahoma Forestry Division State Department Of Agriculture. 9/1981.

Marranca, Bonnie. 1988. American Garden Writing. New York: Penguin Books.

Martin, Laura. 1986. The Wildflower Meadow Book. Charlotte, N. C.: Fast and McMillian Publisher, Inc.

McCoy, Doyle. 1987. Oklahoma Wildflowers.

McKean, A., revised by Mills, R. H. Louisiana Trees. Louisiana Cooperative Extension Service. Pub. 1093, 1/2000.

Miller, George. 1991. Landscaping with Native Plants of Texas and the Southwest. Stillwater, Minn: Voyageur Press Inc.

Mollison, Bill. 1997. Permaculture: A Designers Manual. Australia: Tagari Publications.

Osler, Mirabel. 1989. A Gentle Plea for Chaos. New York: Simon & Schuster.

Reese, William Dean, 1992. Acadiana Flora. Lafayette, La: Center for Louisana Studies University of SW Louisana.

Rice, Elroy. 1985. Pest Control with Nature's Chemicals. Norman, Oklahoma: University of Oklahoma Press.

Sears, Paul. 1980. Deserts on the March. Norman, Oklahoma: University of Oklahoma Press.

Simonds, John. 1978. Earthscape. New York: Van Nostrand Reinhold Co.

Tomanek, G. 1986. Pasture and Range Plants. Hays, Kansas: Fort Hays State University.

Welsh, Douglas; Welch, William; and Duble, Richard. Landscape Water Conservation...Xeriscape. Agricultural Communications, The Texas A & M University System. Pub. B-1584, 7/98.

Pope, Thomas, and Oldenwald, Neil. Shrubs For Louisiana Landscapes - A Handbook. Louisiana Cooperative Extension Service. Pub. 1638, 4/96.

Wasowski, Sally, and Wasowski, Andy. 1997. Native Texas Gardens. Houston, Texas: Gulf Publishing Co.

Index

Cities:

Index of terms and plants by common name:

About the Author

Tom Clote, Jr. who has lived and worked in both Texas and Oklahoma, has worked for over 25 years in the nursery industry. His past experience includes tree seedling cultivation, retail experience with an independent nursery and management for a large chain store. He has operated a design/build landscape firm for 20 years and is owner of a native tree farm.

Mr Clote consults with schools and landowners on native plant rehabilitation efforts. He helped draft and update a City of Norman Oklahoma landscape ordinance in 1999. He serves as the City of Norman's representative on the Viewscape Project, a multi-city wide effort to landscape the Interstate corridors of central Oklahoma.

He teaches landscape design classes at the Moore-Norman Technology Center and is a frequent guest speaker for gardening events.

Mr. Clote has also written A Guidebook to Oklahoma Landscaping published in 1987.

NEED A GIFT?
WE'LL MAIL IT!

FOR ADDITIONAL COPIES OF

Landscape Guide: The South Central States

Send a check or money order for $18.00 for each copy. This price includes postage, taxes, and handling.

Please allow up to 4 weeks for delivery.

Mail to:

Landscape Guides, LLC
P.O. Box 313
Norman, OK 73070-9884

Name: ____

Address: ____

____

City: _________________________________ **Zip Code:** __________

Total number of copies ordered_______________.

Total amount enclosed $____________________.